MANAGING A VIDEO PRODUCTION facility

edited by NEIL HELLER

Knowledge Industry Publications, Inc.
White Plains, NY

T-96

3199 0959

KIPI Bookshelf

Managing a Video Production Facility

Edited by Heller, Neil

ISBN 0-86729-298-9

Printed in the United States of America

Copyright © 1994 by Knowledge Industry Publications, Inc.
701 Westchester Avenue, White Plains, NY 10604

10 9 8 7 6 5 4 3 2 1

Table of Contents

List of Tables and Figures

Introduction

by Neil Heller

The video profession has been affected by the maneuvering in the corporate marketplace. As a result of the economic climate, mergers, acquisitions and consolidations, many companies are looking for ways to cut costs without sacrificing the quality of their products or services. And unless the company is in the business of television, many corporate communications departments that formerly relied on the in-house video production facility are now using alternate means to get their messages across.

Why has the in-house video area been hit hard by cutbacks? The reasons are clear. Video departments are a drain on an organization's budget. They are costly to operate. Investment in equipment, maintenance, supplies and personnel can run high, even in the most modest of facilities. And unless a department can prove that the projects they generate in-house are more than moderately effective in boosting the bottom line, a decision to cut back, or totally eliminate, in-house video production could be a logical and sound business decision.

Another reason for cutbacks has been the virtual explosion of technology that is available in the video production marketplace. Ten years ago, professionals did not have the myriad choices available today. The multitude of formats, ranging from 8mm to digital, is enough to give even the most savvy department manager a headache, as he or she tries to guess which formats will survive in this competitive marketplace. In addition there is the virtual glut of camcorders in homes and offices and the weekend wedding photographer now fancying himself or herself a professional. Years of education and professional experience may not be appreciated when these affordable units are being purchased and used by those in key decision-making positions. When the boss can take his consumer camcorder to the next sales meeting to record the CEO's address, it may cause upper management to think twice about the value of the professional studio and editing suites that are eating up a large portion of the annual budget.

As in-house departments are scaled back, the organization may farm out major video projects to an independent production facility. Many companies are finding that it is less costly to contract with professional facilities for important projects, because these commercial facilities have to keep up with state-of-the-art equipment in order to be competitive. In addition, they are more likely to find a team willing to go those few extra steps, in the hope of garnering repeat business.

THE REASON FOR THIS BOOK

Whether you're presently employed in an in-house video facility within a corporation, educational institution or non-profit agency, whether you're part of a commercial production house or whether you're currently an independent freelancer exploring your options, changes in the way video communications are perceived are very much a part of your future. The downscaling of in-house departments is happening throughout corporate America. And it affects all of us in the video profession.

WHAT THIS BOOK COVERS

Where do you fit in this picture? Do you see your friends pounding the pavement more frequently? What would you do if your department were to be eliminated?

Or perhaps you've always had the ambition to move on to more challenging opportunities. Have you always wanted to be your own boss, make your own hours, run your own business? Are you interested in opening up a full service production facility? Or perhaps you'd like to specialize as an independent, working for various clients on a variety of topics?

This book explores all of these options and provides you with a strong foundation upon which to make some sound decisions on which to base your future in the video industry. Individual chapters cover topics in marketing yourself and marketing your company; analyzing equipment needs and purchasing equipment; writing a business plan, accounting, financial analysis and obtaining financing; and business management and operations. From assessing your needs and desires to conceiving a practical business plan and getting a business off the ground, this book explores the practical aspects of the video business.

WHO ARE YOU AND WHAT DO YOU DO?

Before you decide that it is time to quit your job and go into business for yourself, you need to look at your personal goals and objectives and ask some hard-hitting questions. First, what is your role in your present organization? How do you fit in the overall structure of the company?

Are you a producer, a videographer, an editor, or technician? Perhaps you're a jack-of-all-trades, able to handle all tasks in any given production. Is your present job in a large video facility, where your position is well defined, routine and unlikely to change much over the course of your employment with the company, or are you part of a small shop, where at any one time you might be expected to wear several different hats? Most important, do you like what you're doing?

How long have you been in your present position, and how long did it take you to reach this point in your career? If you went to school for any aspect of video or communications and worked your way up the career ladder, you'll know that competition for the jobs in this industry is fierce. Employers look for the best and brightest. Hours are long. Salaries are low. The work can be exhausting. Yet more and more graduates are opting to enter this field. Why? Because it's fun!

But is the job still fun for you? If you're bored, feeling unchallenged, unappreciated, underpaid and overworked, perhaps it's time to reevaluate where you're headed. How does your current position relate to your long-term personal goals and objectives? Where do you want to be in a year . . . five years . . . ten years? Are you content with the position you have achieved, or are your sights set on something else?

Once you determine what your long term career goals are, take a look at your current position and see what the potential for growth might be. If you're eager to move up, what are some of the obstacles that might prevent this from happening? Are you in a dead-end job, where the only way to move up is to move out?

This book will help you assess your situation and provide you with specific information and ideas to succeed in the video facility marketplace.

PART I
ASSESSING YOURSELF

1 Determining Your Role: Marketing Yourself

by Scott Carlberg

I remember my first day on the job in a corporate TV studio. Not only did I not know where any equipment was, I could hardly find the employee cafeteria. And I was expected to perform video miracles.

From my experience, and from what I have seen of other new producer/directors (p/ds), defining the role of the new p/d is a common problem, even in established facilities. It's a job that has a lot of latitude by virtue of the person who fills the shoes—there are many personal styles of producing and directing. The challenge for the video manager is to help a p/d find his or her style yet fit the needs of the company. The challenge for the p/d is to explore the ways he or she can add something new to the role.

One common problem in new hires: "They often don't do enough listening—listening for expectations." says Kim Krisco, who manages special communications projects for GTE in Dallas. Krisco began his GTE career as a p/d, and eventually managed the corporate video facility. He adds: "People often leap too soon after the job interview and assume they know the complete direction of the job. They revert to old comfortable behavior, or they define the job too narrowly. Either way they don't take enough time to explore new ways to do the job—they fail to be creative."

Krisco's comments remind me of career advice a manager once gave to me. In my situation, the next two people up the ladder were out of town. I had an idea I wanted to try, but I wanted to clear it first. I went to talk with my manager, who was immediately below the vice president of our department on the organizational chart.

My manager invited me into his office. I explained my idea. When I finished, he leaned forward and said, "Carlberg, here's how I want you to work. If you have an idea, plan it out and do it. If it doesn't work, figure out what you did wrong and go on to the next task. If it does work, tell everyone."

His message was clear: You've been hired to think. So think.

SELF STARTERS WELCOMED

My experience as a video supervisor reinforces the wisdom of that manager's advice. I like producer/directors who can "take the bull by the horns" and run with projects. I like p/ds who can take a job description—those vague outlines of what a person should do—and make it come to life. For instance, a job description might say, ". . . creates video programs to meet the needs of our clients." The truly creative video p/d does more than think of a progression of shots, transitions and audio to create a program. The most creative p/ds think about a problem the program can solve—too many company accidents, increasing regulatory fines, poor manufacturing quality, miscommunications between management and employees—and consider that the real project, not just the videotape.

A producer/director who helps clients define their needs and is a problem solver is a business partner for the client.

"Producers are people who *do,* and that's why we're hired." says Barbara Haley, president of

Corporate Media Concepts in Dallas. "Our role is to be a catalyst . . . to create a tangible product from mere thoughts. A good p/d analyzes a problem beyond the audio and video considerations and learns how a video project relates to customer needs, whether they are safety, sales, training or any other corporate communications problem. The audio and video tracks on a videotape are merely tools to help the producer reach the client's objectives."

The Honeymoon

How long does a newcomer have to become an on-target member of the production team? "There's about a four- to six-month honeymoon with a new person," says Kim Krisco. "That's the length of time a person has to get to know the organization. After that, expectations that aren't met will start to 'bite' you."

Where can a producer/director learn about the expectations of the job? The best place to start is the immediate supervisor—the person who did the hiring. This is the person who will probably handle performance ratings, so satisfying the boss is important.

If possible, get to know supervision up the line from the immediate boss. Usually, these people won't supply specifics about the job—such as preferred methods of production—but will provide a broad perspective about the direction of the facility. In an independent production house, naturally, there is a limited number of levels of supervision, which can increase contact with all levels of supervision, making it easier to understand the direction of the facility.

Another source for discerning the direction of the business is from the clients. What do they understand about the medium? What communications problems do they have? How is the customer base changing? For instance, are fewer large-budget product promotion tapes being produced while more lower-budget training tapes take their place? Are more commercial spots on the books? More customer relations tapes? In all these, are clients well-versed in video, or do they need a lot of "hand holding?" The business of the studio provides many cues about the direction of the workplace, and therefore, the best role of the p/d.

Finally, remember that it's easy to get isolated in your own corporate world of video. Don't neglect to maintain your network of video peers.

Use trade associations, seminars and personal friendships to keep in touch with the direction of other video facilities. The current events at other video groups can give you a tip-off on trends that may affect your facility.

Potential for Growth

With the restructuring of business in the 1980s and the continuing corporate emphasis on lean headcounts, all employees of big companies have found that success isn't necessarily defined as continuous movement up the corporate ladder. The organizational structure has flattened. There's less room at the top of the pyramid. Success has to be defined in new ways.

Horizontal growth can take the place of vertical promotions. My latest job is an example of that. Having produced hundreds of video programs and having supervised a facility, there was no higher-rated job in video in my company. The next job I took, management communications coordinator, was a lateral move to the corporate communications group at Phillips Petroleum. I relinquished all supervising duties. I took a new job that had no description, only the identification of a corporate need: help Phillips management improve its communications skills. The job involves speechwriting, article writing, communications consulting, meeting planning and an occasional video production.

This sideways move on the corporate chart expands my communications abilities. It also keeps my work interesting—after 13 years solely in video, it was time for a change. The new job is a way for me to recharge my batteries and develop additional professional skills.

Horizontal moves sometimes go beyond communication specialties. Christy Pearson, manager of executive placement at Foley's, a division of May Company department stores, began at the store as a department manager. She made a series of moves—college recruiter, administrator of video, corporate communications manager, marketing account manager, broadcast producer (for corporate radio and TV spots)—before arriving at her current job.

How did she make these moves? "I've found it's best to constantly assess your strengths and work on your weaknesses," says Pearson, who believes that these moves were no accident. "Position yourself with good organizational contacts. Let people know your qualifications for other jobs in the company.'' Pearson says that communicating her ability to handle television production landed her the video-oriented jobs at Foley's. The corporate contacts helped her find other people-related positions that broaden her horizons and make her more valuable in the corporate staff. In short, the best way to make horizontal moves in a company is to do what comes naturally to video people—communicate.

Vertical growth is the traditional kind of growth—moving up in the organization. There are two ways for this to happen in the corporate video business.

First, people in higher positions in the video department move out of the department and leave job openings for people to fill. This can happen as long as there is enough movement in the rest of the company to let people move to new jobs.

Second, the video organization grows and creates new and better positions. For instance, a video group may grow from three people to 10 over several years, creating a slot for a senior p/d who handles scheduling or a chief engineer who manages an assistant engineer and a technician.

The problem I perceive with the vertical scenario is that there is less control on the part of the individual who wants to grow. Job advancement results from decisions in the corporate hierarchy. For the person who desires vertical growth in the video business, the answer may be to move to a higher position in another video group when learning and growth stop at the present company. This type of movement requires time and energy devoted to self marketing outside the company.

An individual has more control over the horizontal development of his or her career. For instance, the video producer who also is able to publish written material in respected magazines demonstrates versatility to the company and may be opening up job possibilities in related communication fields, such as the corporate communications group.

WHERE VIDEO GROUPS ARE HEADING

The restructuring of business in the 1980s has affected the way individuals regard career potential. It has also changed the face of corporate and independent video production facilities.

Staff groups have suffered tremendous cut-backs as a result of corporate restructuring. As companies looked for ways to cut costs, all but essential parts of the business were trimmed or sometimes eliminated, and video groups often fell under the knife.

Companies are still under pressure to keep quarter-to-quarter earnings up, and that means human resource departments are tracking headcount closely. The result is, at best, slow growth of corporate video groups. Companies don't add video staff capriciously. Companies that never had video but want to have the capability are beginning with one-person shops or contracting for the work with independents or outside production companies.

The trend appears to be contradictory. More companies are using video every day, but the growth of internal video staffs isn't necessarily keeping pace with the use of video in corporations. As in other staff functions, the corporate video jobs are being kept at arms length by keeping the people who do these jobs off the permanent payroll.

If that's the case, what are some of the possible career opportunities in corporate video?

Corporate video coordinator

If the company is only going to have one or two people handling video, one is probably the video coordinator. This is the person who coordinates the independent producers and production houses.

Staff producer/director

These jobs still exist in both areas—corporate and independent houses. In the corporate area, in a lean shop, there may be a sole p/d who is also the technician. As the only real video expert in the company this person may also effectively be the coordinator of all video.

In an independent production house, the p/d may have to produce, direct, and wear lots of other hats, too, including sales. Smaller staffs demand that people be able to handle a larger variety of jobs than groups that allow people to specialize in different parts of the business.

Independents

A freelancer works as an independent contractor and may be hired to manage a complete project or assume one role, such as producer, production assistant or casting director.

The common element in these jobs is *versatility*. In an era when more productivity is being squeezed from every niche of business, the person in the video role can't afford to have limited skills. The most productive and successful video people understand the whole business—writing, directing, producing, editing and marketing the final product. Video itself is a specialized skill, and companies can't generally afford to have someone who is "ultra-specialized" [for example, someone who only runs the computer editor or just works player unit repairs] unless the specialty fills a full-time job by itself and it is cheaper to have the person on the payroll instead of hiring from the outside.

The trend to lean corporate staffs isn't going to stop soon, which indicates that freelance and independent production house opportunities may grow. The challenge for the video professional is to be on the edge of that growth curve in a way that is personally enriching.

CHANGING JOBS

Many video professionals have found personal reward by leaving the corporation and forming their own business. The increase in freelancers is due to two main reasons. First, as corporate video matured as a profession and video professionals grew more adept at their jobs, they saw opportunities to apply their skills outside the parent company. There was perceived to be more monetary and personal freedom by working not as an employee of a company, but working for a variety of end users of the medium. Some video professionals cut loose from the organization and started their own businesses.

That's what Richard Foregger did. He worked for the audio/visual group of a high-tech firm that was acquired by a major company. He sensed that cuts in staff were coming and decided that he would take the initiative himself—he left and formed his own company, America Video. That was in 1985.

Foregger is an independent producer who rented his equipment needs when he started. He then worked through a bank to purchase a beta-cam system. Foregger believes his efforts have provided him with a lot more creativity. Says Foregger, "I'm freer to express my own ideas and

not have them watered down with others. I can pick and choose the jobs I want to bid on. I also like the fact that, as an independent, the financial reward is what I make it—there's no artificial ceiling on my earnings."

But no story is all good news. Foregger warns that the video business has cycles like every other business and surviving financially takes careful planning. Says Foregger, "You have to understand that being an independent makes you ride an emotional and financial roller coaster. You have to develop a thick skin as well as discipline for the times when there is no business coming in."

Some independents have had the choice made for them. As a result of the corporate cutbacks of the past 10 years, many people have had their corporate ties severed for them, and they became independent producers by default.

This was the case with Holland "Dutch" Harpool, who was a vice president of an advertising firm. In a reorganization, he found himself without a job. "I decided to take several months to assess my situation," says Harpool, "but I started getting calls from people who wanted me to do specialized projects for them in the media. Before long, I realized I could make a living being a freelancer."

Harpool's case is unusual in one way—people called him. Harpool had built a good reputation and had lots of contacts from decades in the advertising business. His personal advertising was already done.

In most cases, freelancing means a lot of sales calls before any jobs start coming in. According to Dallas freelancer Barbara Haley, "Having a knowledge of video isn't enough to be a good freelancer. You have to be able to be a good salesperson, too. You have to know how to market yourself."

What other advice do these freelancers have for people who want to make it on their own?

Harpool: "Be prepared to do without corporate support. It wasn't until I had to hire help for special video work or get copies made that I realized the extensive support network that a corporation provided that I now had to handle myself."

Haley: "A good independent is a good businessperson, not just a creative video expert. Besides a good understanding of sales, you need professional financial advisors for accounting and tax help. Don't just figure you'll use a spiral notebook for your accounting or that you'll learn it on the fly. Find a professional to keep you on track."

Harpool: "Understand the market where you plan to work. Find a niche in the market that's not already filled—maybe it's being the producer who can handle low-budget projects for small companies—and start by working that niche."

Haley: "Try not to leave your current job without a client list or something else to bring in money. There's a lot of security in the corporation in the form of a paycheck, insurance, and other benefits. As an independent all this becomes your own responsibility."

INSIDE OR OUTSIDE: A CHOICE

Independent producers extoll the joys of entrepreneuring—forming their own company. It takes a special talent as well as tenaciousness to do that successfully. But it's a challenging and exhilarating experience.

I also think that being an in-company entrepreneur has its own kind of excitement. It depends on what you want from a job. Earlier I mentioned a manager who spurred me to be an independent contractor within the company. What that manager asked me to do was become an entrepreneur in the company . . . in effect, to become an internal freelancer. In the business press, that role has become known as **intrapreneuring.** In video the intrapreneur's job is to find internal business opportunities for video and exploit them for the benefit of the whole organization.

I believe an intrapreneuring viewpoint is a healthy perspective to have in any staff group. It's the best way to open new avenues of programming and customer service.

For instance, the intrapreneur may find that people in the company who may face the cameras in a media event are nervous about it. The solution: a new service—on-camera media training provided by the video group.

The intrapreneur may find that salespeople are having a difficult time getting their tapes shown in remote locations to small groups of customers. The solution: a specially built briefcase with a five-inch TV and an eight millimeter player that run on batteries.

The intrapreneur may discover the unnecessarily high cost of a particular sales meeting being handled by an outside meeting planner. The solution: assume the role of meeting planner, possibly adding some equipment to assist in the job, and save the company the outside fees.

These business situations are the same kind of concerns that freelancers handle for companies. The key is the perspective that the internal producer/director brings to the problem. Does the p/d analyze the situation as an insider, or as an outsider looking in? When the p/d is willing and able to find new ways to solve old problems, that's intrapreneuring.

CONCLUSION

There's a lot more to being a video professional than traditional producing and directing. The real professional understands that video is a means to an end—to solving communications problems, expressing inner talents, or having the kind of creative freedom that video allows—either as an intrapreneur inside or as an independent entrepreneur. In either situation, the successful professional is a problem solver who knows how to leave his or her own stamp on each project while advancing the cause of the client.

2 Choosing a Marketing Strategy

by Scott Carlberg

Railroading in the 1940s is the example many textbooks give of a poor business strategy. The story goes like this: railroads would be more successful companies today if they had realized they were in the business of efficiently transporting people and goods, not in the business of running railcars on steel tracks. Had the managements of railroads considered this definition of their business, they might have been operating some of the best airlines around today, instead of being "also rans" to the aviation industry.

Many video groups are now in their own version of the 1940s. Our railcars are studio productions dubbed onto videocassettes, and it's up to video managers to consider whether current methods of operation will be merely refined or substantially redefined.

In this chapter, we'll look at creating a unified vision of the video group. We'll note how some video professionals analyze their markets and look at the product that video groups really sell.

A SHARED VISION

One person who understands how organizations set goals and define their work is Joe Quigley, a former IBM employee and business consultant. He has worked with the senior managements of many companies to help them focus the reasons their companies exist. Quigley believes it is a leader's job to be a catalyst for development of a shared vision of the organization. He says, "Make certain everyone is pulling in the same direction, toward common goals. The catch is to define the real business of the organization, not just list the activities it performs."

Quigley notes that it's easy to make a laundry list of activities such as "produce high-quality video programs" and call that your business. But these activities represent a method of reaching a goal, not the goal itself.

Says Quigley, "The term 'vision' is important. Vision means something more than actions. It means a philosophy of work, a higher purpose for the actions we undertake. A company's vision and values are the long-term keys to market power and performance."

Companies such as Ford, IBM, Pratt and Lambert, and Brunswick have used this method to focus their business and identify their markets. Usually, after the corporate mission and values are set, individual departments define their business in relation to the corporate mission. Mission statements are equally applicable to the multi-billion dollar company or the enterprise of several hundred thousand dollars.

According to Quigley there are three questions that help a company or department define itself.

- First, what are the organization's basic beliefs? What are the values of the organization.

  Values such as treating one another with respect, maintaining open and honest communications and operating in a safe work environment are examples. Successful organizations aren't afraid to define the way they will work, as well as the work they do. Defining values helps everyone work together.
- Second, why does the organization exist? In other words, what is the mission of the company or department?

  A video group may see its mission as providing visual communications services and advice to meet the sales, training, and communication needs of internal clients at a competitive cost compared to outside suppliers. That's a lot different from "produce tapes," which doesn't signify cost considerations, customer needs or non-tape services.
- Third, what does the video group want to achieve over the long term? In other words, what are the goals?

  For example, a video group may say that it will use its communication tools to save its parent company a dollar amount equal to one-and-a-half times its operating budget yearly through reduced travel and expenses, increased sales, and marketing of internally produced programs to outside clients.

Quigley believes a manager can't do this alone. The manager must harness the ideas of the video staff to write values, a mission and goals that are meaningful to everyone. It can't be done in a vacuum. Notes Quigley:

A manager has a broader overview of the organization than most people, but not the whole picture. Producers have a feel for production opportunities. Engineers know what can be accomplished technically. Sales people or video librarians can articulate the needs of end-users. A manager must draw on these valuable inputs and move the group to a mutual understanding and consensus about the answers to the three basic questions.

In the end, the manager must be the catalyst that forges agreement on a group's mission, values and goals.

A DIFFERENT KIND OF BUSINESS

Two common characteristics of video groups make handling this process a management challenge.

First, video people by nature are self-starters—initiators of ideas, not followers. Opening up a forum about methods of work and direction of the facility will probably open the floodgates to a torrent of different ideas. The manager has to consider them all, because that's the way an innovative vision of the group is created. The manager also has to consider people's feelings and not shut down someone's enthusiasm for the task because an idea wasn't accepted.

Second, video staffs have a dual nature: technical and non-technical. It's been a management challenge since the early days of television. In its most extreme cases, the producer wants to try something the engineer says can't be done, and the engineer doesn't care what's on the screen as long as it has 100% white levels. To bring together these diverse points of view and delineate a common vision for the organization can tax a manager's ability.

Quigley has three suggestions for successfully approaching these problems.

1. *Be up front.* The manager has to be blunt about the reason for creating a mission and values statement. The mission and values statement is an operational tool that funnels resources into the most productive activities for the group. It's a way to communicate dedication to customer service, and it's a way to improve communications among the people in the group. A mission and values statement guides work.

2. *Face criticism.* The manager must also be willing to face criticism during the time the group examines the kind of mission and values it will have. Commonly, employees will use this time to air concerns about company problems, and this is good, because it gets these issues into the open. The manager has an excuse to deal with them but needs an open attitude and tough skin to do it.

3. *Be flexible.* Forging a useful mission and values statement may lead to new ideas for markets and services. A manager has to be willing to consider new ways of doing business or exiting old lines of business. It is important to give careful thought to all ideas, and not to make snap decisions.

Mission Statement Case Study: Maytag Video, Newton, Iowa

The Maytag Co. has used video since the late 1960s when the training department bought a basic "aim and shoot" video system to record and play role-playing in sales training sessions. According to Maytag's Media Services supervisor Jim Jones, the department changed roles due to the demand for different programs. "As time passed, managers began asking for other kinds of tapes—training, employee communication, sales meeting tapes. Soon, 40% of my time was devoted to sales training and 60% to these other kinds of communications."

Video work was steady. In 1986, Maytag hired another person for the facility—now a two-person shop—and Jones realized that the video function was at a crossroads in the company. "I knew that our seven-year-old equipment was worn out. We either had to upgrade and really get in to the video business or get out altogether. I dreaded approaching management with a proposal for all new equipment," says Jones.

The questions he faced were *how* to approach Maytag's management and what the staff would do with the equipment. Says Jones: "I couldn't go to management and say, 'I want new video toys to play with.' That would be the way it looked if I didn't go to management with a good business reason for the purchase. I had seen the way other companies' video groups had published brochures that communicate the kinds of service they provide, and I created a brochure about video for us."

Jones had to redefine the video business of Maytag. (See Figure 2.1.) He did it by borrowing from other companies' video brochures, analyzing the kinds of program requests that clients were making and relying heavily on Maytag's sole purpose of being in business.

The result: "We are more than video," says Jones. "We're here to serve the company—help it make the best washers and dryers on the market—the same kind of goal that the people in the sheet metal shop have. We should either support the goals of the company, or we deserve to be cut from Maytag."

The brochure does more than describe the kinds of service provided by the video group. The brochure spurs the creativity of the clients for new ways to use the facility. It is a catalyst for new ideas.

Part of the way Jones has met corporate needs is by expanding the way clients think about the medium. In fact, he's moved outside the video medium to keep his staff in-tune with company needs. "We are called 'Media Services,' which is more than video," says Jones. "We help with meeting planning in addition to video, because that's where the company needs our help." Jones has developed broader communication skills in order to keep his promise to help the company produce a high-quality product. He's analyzed the market and filled the needs.

Jones had an open house to display the studio renovation. He worked through the secretaries of Maytag's top management to schedule a time they could attend the open house. Top management got quick demonstrations of new video capabilities and testimonials to the way the studio helps Maytag meet corporate goals. Jones' reward: "Top management gives us more work than we can do. We have supported the goals management has set for the company, and it supports our media group."

UNDERSTANDING YOUR INDUSTRY

Writing a mission and values statement for an independent organization is also useful. In order to write it, there has to be an understanding of the industry and of the market.

My best example of understanding a business isn't a video example, it's a photography example. A friend of mine wanted to make his living as a photographer, but he was unknown in the Colorado town where he took up residence.

In was winter. He was poor and looking for clients. He noticed in the local newspaper that a dogsled race was going to be held near his city that weekend. He went, tripod and camera in hand.

He positioned himself at a critical turn in the race. He shot each musher one by one as they daringly took the curve. When all the racers had

Figure 2.1: Maytag Video Definition

—We support the goals of the company through efficient, cost-effective communication.

—We are a department of professional communicators.

—Our purpose is to take your ideas and messages and communicate them to your audience.

—Our primary service is videotape production, but we can also arrange for talent, music services and media equipment rental.

—Our video production facilities offer fast turnaround, lower cost, convenience, quality, and capabilities equal to or greater than outside facilities.

—We have a staff experienced in all aspects of video production.

—Our staff is familiar with Maytag products and procedures.

---

passed, he got a list of each sled number and the home address of each racer.

He developed the pictures and sent each racer a portrait of their derring-do on the curve. The proof he sent purposefully had a line through the middle of the shot (done in the darkroom), which allowed the racer to see the shot, but made the print unworthy of display. A price list for enlargements of good shots was included.

The racers didn't even see him shooting the pictures, nor did they know they wanted a picture of themselves and their dogs until the photographer contacted them. My friend sold many pictures.

What market analysis!

What follows is a story of how similar market analysis can be applied to the video medium. When "Dutch" Harpool became a media freelancer after several decades in the advertising business, he realized he had to define his business and communicate his line of work to potential clients. He is a writer, producer and director. His business card said, "Holland 'Dutch' Harpool, Specialist in Communication and Persuasion."

"I do more than write produce and direct," says Harpool. "I solve communication problems. When people use my service, they buy more than the media abilities I have. They also buy trust, dependability, and my analysis of their problem. When I approach a client, I ask myself, 'What's the problem that exists for the client, and what can I do to solve it?' That only makes sense. Can you imagine approaching a client by asking, 'I wonder what sort of special effects I can use.' It wouldn't work. That's choosing your tools before you know what kind of project you're working on."

Harpool knew from his advertising experience that he wouldn't do well as a plain-vanilla freelancer. He needed to add something special for the client. Many freelancers and independent companies find that's true. Says Harpool, "I did several things to build my business. First, I looked for a niche that wasn't being filled in order to get a toe-hold in the market—something that I could use to build a reputation. I found this niche in my ability to handle low-budget, quick turnaround video and advertising projects.

"Next, I built a network of agencies and production houses. I told them I was available to pick up the overload on projects they had, or they could refer the projects to me that they considered too small to handle. I told them I would refer large projects to them. This did two things. It solved their problem of turning away clients or finding clients, and it built goodwill between us.

"Finally, I determined the way I would bill my clients. I settled on a per project basis, not a per hour basis. I found that my market wanted a definite idea of costs at the front of the project. They wouldn't get that with an hourly rate, which is the way many freelancers work. My project rate is firm—predictable in their budget—and my clients like that."

Harpool says that he learned about the local market by getting out and talking with people. He keeps in touch with ad agency people, production house staffs, and other freelancers to get a feel for the market.

The same sort of research applies to independent production houses and even to corporate video groups. Every video "group"—whether a one-person business or a video facility—has to know the likes, dislikes and needs of the local market to sell its service.

"Niches are important," according to Barbara Haley, president of Corporate Media Concepts. "The most successful independent producer or

Market Analysis Case Study: UNIVISION, Syracuse, New York

In 1976, Bob Romano read a *Newsweek* article about wide-screen TV projectors. He thought that people would want something like a wide-screen television system—he was certainly interested.

So he traveled from New York to Boston, to the headquarters of the TV projector company, and told them that he wanted to sell their TV projection systems. He got a job with a dealer.

In 1977, after outselling all other upstate New York video projector salespeople, he was ready for his own dealership. Romano started with a one-projector line of credit from the manufacturer and a Chevy van of his own. UNIVISION was born. He developed that idea into a business that sells more than $10 million in video hardware and services annually.

UNIVISION build a projector sales business by tapping the "entertainment" market: bars and hotels. "People were enthusiastic about wide screen TV," says Romano. "And I could see that a video revolution was coming. I added a line of half-inch players, then cameras. I began marketing to business clients, not bars. Businesses were more open to new ideas and needed solutions to their communication problems. I helped answer them."

As the business grew, Romano added staff. Romano realized that he would have an internal communications problem if his employees didn't understand the nature of his business. "We've had a statement of objectives for UNIVISION since day one," says Romano. (See Figure 2.2: Objectives of UNIVISION Video Systems.) While the UNIVISION goals are longer than many, they are right on target for describing the purpose of the company and methods of work.

Romano's experience showed UNIVISION has to be a dynamic organization, not a static company. As product lines change, so do the capabilities that UNIVISION delivers to its clients. The corporate objectives must allow for flexibility.

For instance, Romano has seen significant shifts in business since the company began in 1977. What began as a small hardware sales firm with one product line grew to include several product lines. As complex video systems were sold, customers wanted training in program production. UNIVISION provided training for video users, then sold the equipment to produce programs.

Eventually, clients wanted a completely new kind of service from UNIVISION: producers and directors. Romano added a production service, which was spun-off into "Axxess," an independent production company. Why were the two entities separated? "The demands for a specialized production professional are different from those of a professional hardware salesperson," says Romano. "The two aren't compatible in one company. Clients would call and want to talk production and hardware people would answer the phone, or it would work the other way. We needed to separate them to properly serve the client." Axxess is still evolving and Romano is considering the way a statement of company objectives fits that business.

What else does a written statement of objectives do for UNIVISION? "It keeps work interesting and productive for my work associates," says Romano. "Our objectives mention self development and expanding our knowledge, and our employees take that seriously. I invite employees to create their own job write-ups and set goals on the kind of jobs they want." This sort of initiative eases the everyday personnel management Romano has to do—people direct their own work. Romano says that coupling the corporate objectives and employees' creativity on job write-ups results in innovative approaches to problems in the business.

production company identifies a niche before the market pinpoints it as a need [shoots the dogsleds during the race], and stakes the territory early. An independent can become known as the quint-essential audio/visual meeting planner for a metro area before anyone identifies the market. Or, a company can be the first to make video confer-

encing easy in a market, or become the expert in creating video annual reports.

"My niche was built as one of the first independents to cater to the nonbroadcast market. I was the person whom companies could call and be assured that I knew the corporate angle for a videotape. I wasn't a traditional 'broadcast

Figure 2.2: Objectives of UNIVISION Video Systems

Objective I: To be a progressive, aggressive, quality growing video communications firm.

Objective II: To offer our clients continuing and outstanding service in solving their video communications requirements through the adequate and proper use of our products and services.

To try in the light of all the circumstances surrounding our clients, which we shall make every conscientious effort to ascertain and understand, to give them that service which, had we been in the same circumstances, we would have wanted applied in our own business.

Objective III: To recognize the benefits of continued self-development of each video specialist of this corporation. We are individually obligated to continually expand our knowledge, experience and skills for the benefit of those it is our opportunity to serve.

Objective IV: To provide each associate with the services, facilities and business atmosphere which will enable him to achieve his personal goals as a member of our organization. To conduct our individual business activities with consideration for the others associated with us.

Objective V: To recognize the importance and necessity for the growth of our organization individually and collectively.

As individuals, we will only achieve real satisfaction by constantly seeking and experiencing a greater utilization of our full potential for success in our careers—no matter our present level of success. This recognizes that some of us must work harder and some must learn to work smarter.

To recognize that the expansion of our corporation through the addition of carefully selected and properly developed consultants will benefit us as individuals and that is a necessity if our corporation and our industry are not only to grow in importance and stature, but also merely to maintain our relative position in the economy of which we are a part. By careful selection, to bring into the corporation only those who measure up to high standards and are personally acceptable to our present associates.

Objective VI: Through mutual understanding and respect for each other's duties and responsibilities—a cordial relationship between office and staff and the video specialists be maintained so that we may work together as a professional organization.

Objective VII: To recognize that we are an association of individuals who are primarily in business for ourselves. In order for such an association to be a happy and beneficial one, each of us must recognize and discharge our responsibilities and obligations to our clients and to each other and the company we represent. Despite the individual opportunities and nature of our association, we recognize the satisfactions and benefits that result by our working together as an effective and proud team—a winning team that is continually and increasingly gaining the respect of our clients, our competitors, our associates throughout our company and, more importantly, ourselves for the job we are doing and the way we are doing it and the satisfaction and rewards resulting from it.

---

transplant.' That was how I got my foot in the door."

There are several good ways to get a foot in the door.

1. *Introduce a new trend.* Try something that hasn't been commonly used in your market. One producer keeps in touch with markets similar to his, talking with video professionals who have similar situations, but wouldn't be competing with him in his market. They share programming ideas.

2. *Take a contrary view.* Some independents look where the herd is going and then head in the opposite direction. For example, if everyone in the market is going with high-end computer animation, there may well be customers left in the

cold who have modest computer animation needs. This could be a niche to fill.

3. *Find a specialty.* Barbara Haley became known as the producer who understood the corporate viewpoint—could organize a disorganized project with lots of details and little time. Some independents have specialized in computer animation, training tapes, management education, quality training, environmental engineering, and more. Specialties depend on what trends businesses see affecting their balance sheet.

4. *Ride your reputation.* This is chancy and good for only a limited time. If you have built a high-quality reputation at your current company, you can use that to a limited extent as a springboard in a freelance career. However, you have to move off the starting block fast, because when you leave the place where you had the good exposure to the market, people can forget about you in a hurry.

THE CHALLENGE OF THE MARKETPLACE

This is a challenge: innovation—to be ahead of the competition. "People who start their own business can make a big mistake," says Barbara Haley. "They may see an independent hitting it big in his or her niche, and think, 'That's what I want to do.' Then they try to exploit the same angle. The problem is that the person they're emulating has already filled that niche—built a name—and has the market fairly well captured."

How have other people "made it?"

- One independent company marketed video as a sales tool for realtors. The company produced a program promoting the city as a good place to live and locate a business. Realtors sent copies of the program out to their affiliated offices nationwide. This market was local.
- One producer was the only person in his Midwest region with a Steadicam, and everyone called him because of this marketing angle. This market was regional.
- One company in a vacation spot became known as the company that had built a network of travel agents and travelers, so they

specialized in marketing travel promotion tapes. This market covered at least two-thirds of the country (those people within easy travel to the destination.)

- One company built a tape publishing business in petroleum training by selling geological and geophysical training-tape concepts to a consortium of oil companies that financially backed the productions for a set number of years at a yearly subscription rate. The oil companies, having cut training staff during a drop in oil prices, found this to be an inexpensive way to provide some training. This market was worldwide.

What are common ways that video groups mis-read or mis-handle a market?

- No analysis of communication problems. Freelancers, corporate producers or independent production houses let video stand at face value and figure clients will buy it for what it is. In this case, the producer is running railcars, not providing transportation.
- Weak marketing. One producer told me that he had done a great tape for a local hospital and had determined that many other hospitals could use the same kind of tape. (And they probably could.) So he sent letters to a long list of hospitals proposing the same kind of tape for each one. He received no replies. His letters offered nothing special—they were part of the mail delivered everyday. Nothing made his pitch different than the others that arrived in the mailbag. A personal, well-researched call could have made a big difference.
- Inflexibility. Production houses that fail to change with the times suffer the consequences. Imagine where Bob Romano would be today if he decided he would sell only his original TV projection systems. Find the direction of client needs and be on the leading edge.

INTEGRITY ABOVE ALL

For all of the different marketing strategies, there's really one characteristic that freelancers, independent production houses and corporate video facilities strive for . . . integrity.

All of the scripts, the casting calls, shooting and editing come down to the character of the people involved in the program. The best program on tape does the client little good if the people behind it aren't ethical, caring and open. That's part of the vision of the organization. That's what keeps clients coming back. It's the one market niche everyone can share.

3 Finding a Marketing Niche: Three Case Studies

by Scott Carlberg

Most new businesses fail within five years of start-up. What do the survivors share that makes them successful?

Here are three case studies in media production businesses—a small, a medium and a large operation—all veterans of the start-up. While each entrepreneur created a distinctly different operation, several business practices are common to all three.

1. Each owner/manager has a focus for the business. He has defined the clients he'll serve and works hard to develop that niche.

2. Each operation spends money judiciously. Financial decisions are meticulously analyzed for their effect on profits and cash flow.

3. Each business started small. The entrepreneurs had realistic expectations of business, built the operation in steps and continue that pattern to maximize future success.

VIDEO FAMILY AFFAIR: ALPHA VIDEO PRODUCTIONS, DALLAS

Alpha Video Productions in Dallas is truly a "mom and pop" affair. Gary and Susan Bauer moved from Los Angeles, where Gary did freelance video production work, back home to Dallas in 1980. They started Alpha Video.

Their principal assets were time and talent. They had no equipment. "I rented equipment packages to do what I had to do," says Gary. When his rentals were frequent enough to prove the continued customer demand for a service and justify payments on his own hardware, he bought equipment and slowly pieced together a production operation.

Indeed, he created his business by building program pieces for his clients. Gary's earliest editing in Dallas was for ESPN and CBS Sports, building the bodies of sports segments for network use. He had no grand plans to produce elaborate tapes. His goal: create billing.

Bauer's method was to knock on production house doors. He freelanced as a cameraman/producer/director. Slowly, his reputation as a reliable producer spread around the Dallas market, and the doors opened for more freelance business. "The key to my success was networking," says Bauer. "I got around to meet a lot of people. For instance, one production house producer told me about a local doctor who had video equipment and wanted to produce programs, but needed help. I called on the doctor and produced the programs." Even unconventional contacts helped. "I maintained good communications with equipment vendors," says Bauer. "They'd let me know who bought hardware and could use production help."

This differed from Bauer's original plans for making a living in Dallas. "Originally I thought I'd freelance for a while, then settle into a job with one production house. As time went on, Susan and I found that we were building up enough work to start our own business. That's what we decided to do in 1981."

In time, Gary developed a strategy in the Dallas teleproduction market—find a niche and work it. "While I occasionally do complete productions

17

for medium and large corporations and still do short network packages, my business has specialized in clean, low-budget productions for charitable groups. That doesn't mean I give the programs away. I have learned how to produce a good fundraising tape on a tight budget."

Bauer has a corner of the video market in Dallas. He welcomes the kind of clients other production companies don't actively pursue, and he makes a living at it.

Gary and Susan are the only two employees of Alpha Video. Susan keeps the books for the business. "It has to be a family affair at this level," says Gary. "It's full-time for both of us."

They steadily increased billings. From first year billings of $23,000 (1981) to $100,000 (1984) to $141,000 (1987). In 1985, the production load warranted an addition to the business family, and the Bauers began to use a production assistant whose previous experience was only a short time in a TV station. Bauer was able to give the production assistant an increase in pay over the broadcast job, starting him as contract labor. Later, the production assistant became a full-time employee. Bauer used local contacts to survey the going market rates for a production assistant as well as trade association salary studies. Bauer found that with his low-overhead shop, monthly billings of $15,000 would warrant use of a regular production assistant.

In 1988, the financial success changed. Dallas was hit hard by the depression in the oil business as well as the effects of corporate restructuring. Alpha Video's billings dropped by a third. "It was horrendous," says Bauer. "But on the positive side, we had low overhead costs—not a lot of equipment, a small office, only one employee besides Susan and me—so we could fare better than someone with high fixed costs."

The Bauers had seen the decline coming. Susan tracked the cash closely and detected the impending financial crunch. Alpha Video dug-in for the fight. "A lot of clients went out of business or stopped using video," says Bauer. The lesson in this is that "tracking" is more than a video term; it's important to track finances carefully even in the smallest company in order to detect future money problems.

The business slump in Dallas had an effect on the whole local video production industry. "A lot of people had entered the Dallas video market when it was doing well," says Bauer. "The crash weeded out people with too much overhead—

people who were dependent on the good times." That made life a little easier for the survivors.

During the financial crunch, Alpha Video's production assistant decided to set-up his own freelance shop, which probably saved Gary from making a tough personnel decision. The production assistant's decision also saved Alpha Video some financial difficulties.

"We haven't hired a new production assistant," says Gary. "I'm not ready to commit to another person now. I need to see continued long-term strength in the production market here first, then hire a good all-around production helper."

That philosophy is maintained by Bauer even in making equipment decisions—make each dollar spent on equipment do the most work possible. For instance, when Bauer needed a new camera, he analyzed the requests of his clients: a demand for chip camera benefits and a choice of raw footage formats. Bauer bought a SONY-M7 camera, a chip camera, that lets him deliver footage on 3/4-inch or beta tape stock, whichever the client wants. (They're demanding beta four-to-one over 3/4-inch.)

Bauer stays with a bread-and-butter operation: camera, recorder and a cuts-only edit system. "I can go around the corner in Dallas and get all the effects I need," he says. "So I don't need to invest my money in a lot of equipment for special effects and have the pressure of meeting financing costs each month in order to break even. My client base isn't oriented to fancy visuals."

While no production company can afford to make the wrong business decision, a small company has the least amount of room for error. "The amount of available cash from operations—that's the problem of being a small business," says Bauer. "I find that renting the equipment consistently two days per month for six to eight months, usually at one to two percent of the equipment purchase price, justifies the purchase of the equipment for Alpha Video. That tells me two things: first, there's a continuing demand for the hardware, and second, I can finance the purchase at that rate of rental activity."

For example, a SONY BVW-35 deck would rent at $250 per day in Dallas, which is between one and two percent of its $13,000 list price. At two rental days per month, that's a $500 monthly rental charge. Financing at ten percent for a 60-month period (five years) would cost $275.19 per month. This assumes that the vendor charges the list price. The monthly charge can be lower if good negotiations result in a lower initial purchase price.

Rentals let Bauer establish the way that a new piece of equipment dovetails into his business (if at all) and let him be certain that the hardware will be in demand consistently—not just as a fad in video electronics. Says Bauer, "I prefer to let the market determine the long-term need of the equipment."

When Bauer buys equipment, he usually works through a local lender for financing. His best working relationship has been with his credit union. He uses equipment lease/purchase plans when the tax laws work to his advantage—being able to write-off lease costs versus take lease costs as expenses, which come off the profit/loss statement. Bauer works closely with his CPA to determine the cheapest money to purchase each major piece of gear.

Perhaps Alpha Video's biggest challenge is technical back-up for his equipment. Bauer isn't an electronic engineer—he's a producer/director. He tackles that problem in several ways.

First, Bauer uses local vendors who can back-up their equipment. He needs quick response time on equipment problems.

Second, Bauer ensures long-term equipment care through a contract with a video engineer who flies into Dallas to carry-out a regular maintenance schedule: major overhauls of equipment (replace heads, change reels tables) every year; tune the system (new brake bands, springs) every six months. This routine works well for Alpha. Emergency maintenance trips are billed at a pre-established rate in the contract.

Third, Bauer does basic maintenance (head cleaning, change out certain belts, reset basic levels) himself.

Alpha Video started and remains a small business. That has been by design. "I enjoy running a business this way," says Bauer. "I had some decisions to make. I decided that my family is important to me, and I allocated a certain amount of time to my family. That means that I turn away some business or use several freelancers to get certain footage for me. The result is that Alpha Video may not bill as much as it could. On the other hand, Susan and I make a living with a solid base of clients, and we have guaranteed that we have all the things in our lives that are important to us."

What are the plans for the future? "I have several decisions to make," says Bauer.

"My lease comes up for renewal this year. I need to decide whether to renew for a year, three years or five. Obviously, the lowest monthly cost is for five years." There's another consideration in the lease decision—whether to renew the lease at all. The Bauer's are considering whether to go "on the road" on a long-term overseas trip to produce video programs for his church clients.

Though the "stay in Dallas or sell the business" choice looms, Bauer still has other decisions to make. He has equipment concerns. Alpha has to consider a new edit system. The cuts-only 3/4-inch system may give way to an A/B roll Betacam system to meet increasing customer demands for this kind of work. "The economy is coming back, and buying an A/B system may be more economical then renting," says Bauer. "I want to be ready with the right kind of hardware—not wait for delivery, debugging and getting accustomed to the system."

From a programming perspective, Bauer is working his niche. He recently presented an introduction to video to a national convention of church-affiliated colleges—principal clients of his—in an effort to get Alpha Video's name before the public. Bauer already enjoys an outstanding record of successful fundraising and recruiting tapes with church colleges.

Bauer also says he needs to make plans for hiring a production assistant if business continues to increase. "I'll look for a recent college grad or intern to help in operations. I can teach many of the basics of video operations, but I definitely want someone with good people skills. That can have a big affect on this business."

Success isn't always measured by the growth of a company. Some video entrepreneurs have started a small business and decided to be successful as the owners of small businesses. Their success is measured in a stable client base, the responsiveness to changes in the market, maintaining their proper time for family and the excitement of being at the helm of a desktop business.

TRANSFERRING BUSINESS SKILLS: APPLE TREE PRODUCTIONS, BOISE

Not everyone who is now in the video business has always been a communications professional. Calvin Cotton, president and major stockholder of Apple Tree Productions in Boise, Idaho, proves that.

Cotton finished high school through a high school equivalency correspondence course and has some community college background. For 14 years, he serviced consumer electronics, mostly TVs.

Then Cotton sold chemicals, an experience he says was his transition from the consumer to the industrial side of business. "It's the side of business I prefer," says Cotton. "When I was selling chemicals, I learned how to spot customer problems and find solutions for them. I also learned a lot about human nature and the way to deal with organizations at all levels." It was important to Cotton that his success was easily measurable in chemical sales—he could tell how he was doing—because, he says, "My checkbook was my order pad." If Cotton sold chemicals successfully, his checkbook balance reflected it.

Calvin Cotton was most well-known in Boise for being the chief executive officer of Timberline Wood Stoves, a highly successful company. It was a business he built from scratch with a friend. In time, he decided to sell part of the company to his partner. He wanted to try something new. Cotton did some business consulting—helping another stove company diversify into year-round products. Then he returned to Boise.

"I looked around Boise to see what was needed," says Cotton. "I was aware of the growing need for corporate video, but in Boise, there was no permanent video facility for companies to use." TV stations had been serving corporate clients, but given the daily operations of broadcasting, they couldn't offer the kind of production dedication to the industrial side of video that Cotton thought was proper for business.

Cotton had used video in the stove business. He wrote his own commercials (award-winners) and helped the production crews produce them.

Cotton spent six months in the library studying about video, then turned some hard assets liquid and, as Cotton says, "bought what I could afford"—a cuts-only 3/4-inch edit system. Cotton set-up his edit suite in a spare bedroom of his home. "I made calls in person to potential clients, just like in the chemical sales business, and told them about the video services I could provide," says Cotton. At this point, Apple Tree started an "awareness campaign" about its existence.

Apple Tree has been part of the Boise business community since 1981. It was the first quality independent production facility with staying power in Boise.

It wasn't easy. Says Cotton, "I was still known as the 'Stove Baron,' not someone who runs a video business." His in-person calls (done consistently) and dedication to quality showed the local business community that Apple Tree was serious about Boise.

The stove business helped Cotton bring something to Apple Tree that many video operators lack: business experience. "I had something that a lot of video operators don't—a background in customer service, electronics and running a business," says Cotton. He had the business savvy, but not the video production background. "I became a student," says Cotton. "And when I became proficient and developed more business, I brought other people into the business."

The original kind of productions that built a base for Apple Tree remains the base of productions today—internal and external corporate communications. For example:

• Apple Tree has been riding the wave of corporate quality-control programs. An Idaho food company has Apple Tree taping its monthly "Total Quality Management" meetings for distribution companywide.

• Corporate promotional tapes have been a money-maker for Apple Tree. A conglomerate with an eye to overseas markets had its overview tape re-edited and made in Japanese, Korean and Mandarin versions.

• Training has its place in the Apple Tree mix. A regional bank commissioned Apple Tree for its teller training.

• Daily operations have also played a key role in the success of Apple Tree Productions. Seemingly mundane activities such as duplication of tapes and bumps from one format to another generate cash flow. A dollar is a dollar no matter how it gets into the company.

"There are a lot of top companies with offices in Boise—Hewlett-Packard, Albertson's, MK, Simplot, Boise-Cascade, Ore-Ida—and all of them need local video work. It's convenient for them to work locally," says Cotton. That fits with what many corporate video managers have experienced in servicing field locations—with lean staffs, producing every requested field production is a difficult task. No wonder some offices away from headquarters turn to production facilities for help.

That's just the market Apple Tree fills. "There were lots of former broadcasters doing freelance corporate video work in the region," says Cotton. "So I started an edit house—somewhere that these freelancers could go to finish what they shot. I found editing didn't completely support itself, so I hired someone in sales and made this a full service house."

But working with local offices isn't the only business Apple Tree does. "We have clients from New York to Hawaii," says Cotton. "I think there are several reasons. We produce quality, and quality always has a market. We have good rates. And Boise is an awfully good place to come to work for a while."

The shop recently increased staff to seven employees: two producer/salespeople, three production assistants, two secretaries, a general manager, and the president/chief editor/director—Calvin Cotton. In the shop, his title is "Cal."

Revenues have risen. Apple Tree's first year grossed $50,000. In 1989, Apple Tree grossed half a million dollars. Business has increased by "good word of mouth," according to Cotton, and direct mail campaigns to area offices, followed-up by one-on-one sales calls. "The marketing people keep busy with new ideas to demonstrate how video can be used in business," says Cotton.

For instance, some additional business for Apple Tree in 1989 resulted from an open house held at the facility. Regular clients were invited, as were potential customers. Invitations were sent to businesses and entrepreneurs big enough to use video profitably, but who weren't using Apple Tree currently. People flowed into the facility, partly because they wanted to see Boise's quality production facility. (As many video managers know, the public has a natural fascination with the TV business.) Once in the building, the Apple Tree staff had the chance to demonstrate their capabilities. A video wall showed recent productions, editors staffed the booths, and cameras, dollies and sets were out. "We fed our guests food and information," says Cotton.

Direct mailers were another sales approach. Hospitals in the region were targeted with a letter, brochure and mail-back card. Medical administrators could check off and mail in their request for medical program information including topics such as post-operative care, advertising special medical services or producing hospital newsletters on video. Cotton's producers act as any independent producer or corporate producer upon a program request. They confer with the client, prepare a budget and create the tape.

The sales staff applies basic sales techniques to video—handling sales obstacles, creating product awareness, going "door to door"—all techniques that might be associated with selling an appliance, not a communications tool. Apple Tree people know how to sell, which is a weakness in many other video facilities.

"We fill business needs," says Cotton. "Apple Tree is the in-house video staff that companies don't have at their office. We produce the communications tools they need in order to do their jobs. Our producer/salespeople show up on the client's doorstep and make it easy for clients to get the communications tools they need."

One sales technique, keeping regular contact with customers, is particularly important to Apple Tree. "There's always a large turnover in corporate offices. We have to keep in touch to know who's running the department, so businesses know we're here to serve their communications needs."

However, Cotton says the job of selling isn't as tough with video as with some other products. "Industry is interested in video. Industry is already using video. So selling our services is a matter of helping their communication needs come to fruition through video."

Usually that means Apple Tree has to supply a well-produced, clean program, not something with a lot of "technical garbage," in Cotton's terms. "There's a lot of video clutter in many programs." says Cotton. "The trap is that some people want to twist a picture rather than deliver a message. A bad editor can overuse electronic effects as a crutch—an excuse—for good editing."

This is more than a consideration of production style. It's a survival strategy, too. Says Cotton, "There's a high fatality rate in this business. I decided that Apple Tree would 'grow as we go' and not try to have every goodie in the hardware market."

Cotton says that even so, cost of equipment is a problem. "We have to service a lot of clients to stay financially healthy, because we have had to invest a lot of money just to stay at the state of the art, even to stick with 'bread and butter' products." Cotton uses two criteria for equipment purchases. First, what's necessary to get the quality we need? Second, can Apple Tree realistically afford the equipment? These must be in balance with the overall customer demands.

What does the future hold for Apple Tree? "We're not going to expand the current facility. We'll work on making this one more efficient," says Cotton. "Our emphasis now is creating new ideas for communications tools for manufacturers."

With his roots firmly in Boise, Cotton plans to branch out into even more innovative formats of video programs.

"NICHE-MANSHIP": COOK SOUND AND PICTURE WORKS, HOUSTON

A production company doesn't have to offer every audio and video service to be a big company. Cook Sound and Pictures Works of Houston made its niche a big affair.

Dwight Cook began his communications career in radio. He worked both sides of the microphone for 10 years, principally as a production director. The fast-paced radio production environment accentuated some differences between what many clients wanted and what the typical radio station could deliver on a tight time and cost budget.

Clients wanted higher production values in sound and asked Cook to spend his own time and talent to produce the product they sought. In 1975, after spending more and more time improving audio values for clients, Cook decided he could make a living at audio production alone. He left his job at the radio station.

"I wanted to be more responsible for my future," says Cook. "I had developed my list of customers in audio and saw a chance to make a break."

The first studio Cook used was set-up in his home. The equipment was modest: a couple of multi-track audio recorders and microphones. Cook says his first audio equipment was chosen on the basis of what manufacturer would give a start-up business an account to buy hardware.

The home studio was "in production" for eight months. Cook's next office was in a small street-side office complex. This location underwent three expansions—an audio booth, a studio and more office space. The company stayed in this location for seven years.

In 1984, Cook moved to a more prestigious business location, the Galleria complex of southwest Houston. "Location has a lot to do with our success," says Cook. "People want to do their work in a solid business-like environment." His company's business grew yearly at an average 25 % rate.

Cook's analysis of the audio production business was correct—there was, and is, a growing market for high-quality audio. "People expect more from audio now than they did in the past," says Cook. "In electronics stores, the home VCRs have high-fi sound, most TVs sold today are stereo and compact disks have changed the way people listen to music." Cook says his services help his clients meet this heightened awareness of audio for their clients who are the end users of the audio track.

Obviously, the emphasis of Cook Sound and Picture is sound. "We don't do typical productions—a needle-drop of music, for example. We do everything as a *sound painting*, the whole feeling that sound creates. It may take the blending of several pieces of music, music and sound effects, layering effects to create new sounds, changing the background or ambience of a soundtrack or recording new music and sounds." Clients may come to Cook to make someone sound as though he is on-location at a basketball game in a huge arena, to recreate a new piece of orchestral music or to add consistency to the ambience of a film-style two-person location interview sequence.

But the word "picture" is part of the company name, too. "That's an accommodation, not a focus in our business," says Cook. "We had a natural evolution into video with the multi-image soundtracks, commercials and corporate videotapes that came in as a part of our audio business." It was six years ago that Cook added video capabilities—an in-house edit bay—that was mainly geared to enhancing the company's ability to match audio to video.

Cook's company will coordinate the freelance crews and do the editing on a videotape if it's part of an overall project, but audio is still the kingpin at Cook Sound and Picture Works. "The word 'niche' is the key," says Cook. "By sticking to my niche—audio recording, mixing and enhancing—I'm better able to serve my clients. I can do more for the client's dollar by sticking to what I do best."

Among Cook's clients:

• A major computer company used Cook Sound to produce the soundtrack for a videotape that would show in a specially built mini-theater at a national trade show. The tape was mostly music with almost no talk.

• An American auto manufacturer hired Cook to make the soundtracks—voice overs, effects and original music—for six commercials.

• Regional amusement parks have hired Cook to capture the unusual sounds of their rides and attractions for use in TV commercials.

• Corporations have used Cook to mix final tracks of audio and videotapes for corporate communications.

• Various clients have used Cook to duplicate audiocassettes. In the past year Cook has done over 200,000 audio dubs.

Despite the large size of his business, Cook considers his niche as a "boutique" concept of production, and he sees this concept as a trend in the media business. "Production work is going to get like the medical field—a field of specialties. Some people try to do everything themselves, and usually something will suffer when they do. There's pressure to make the requirements of a client's project fit the mold of the people on your staff, so instead of hiring a freelancer who may be the best person for the job, management settles for the abilities of the person in-house."

Cook's staff reflects that philosophy. While the company does regional business with corporate communicators and national business with film producers and ad agencies, only eight people make up the staff of Cook Sound and Picture Works: an accountant, office person, marketing manager, duplication operator, studio manager and three engineers. Cook also does some of the marketing work. Cook uses contract labor and interns to fill other jobs, as needed. He custom-builds the best crew of freelancers for each project, and his permanent employees coordinate the work.

This strategy saves more than the payroll costs of keeping permanent employees. It saves capital costs, too. Since Cook emphasizes audio, he invests in the necessary audio hardware for his specialty. Since video is a capability, but not an emphasis, he doesn't have video production hardware. "We provide online services with our own 3/4-inch-SP and one-inch, but we don't have equipment to shoot," says Cook. "There are other people in the region who can do that." Cook reads his market well when he says that a number of people are making a moderate living in the production business, while he can be successful with a specialty in sound. There's no business reason to enter the video production fray with a lot of equipment.

Cook isn't closed-minded about new ways to serve his clients, but he does guard his equipment dollars. Like other production houses, he prefers to spend money renting hardware if he's not certain the service a piece of hardware provides has a long-term place in his facility. If demand proves tenacious, and the service fits with the marketing strategy of his company, Cook will consider purchase of the equipment.

One market that Cook serves is the small production shop. Small producers can successfully run a small production service, without investing in lots of expensive equipment, because fancy production effects are just around the corner in many markets. Cook is the fancy production house for audio, and part of Cook's revenue comes from small media operations that need occasional help in audio.

Cook as tried several avenues to market his operations. "Word of mouth has been our best advertising," says Cook. "We tried other methods of publicity. The advertising pages of the phone book didn't help at all. We deal with businesses, not the general public, and the general public looks at the phone book."

"We tried display ads in video magazines. That didn't help because our peers read it, not our potential customers. Our customers are ad agencies, film producers and corporate communications groups. We tried trade magazines in the advertising industry and got some response."

"We've had a lot of success with direct mail campaigns and with our newsletter. We send these to current clients and specialty lists of potential clients in the businesses we serve." Cook's newsletter promotes new services of the shop, introduces new employees and their specialties, explains technical audio concerns and inventories new sound effects and the ways they can be used in productions. The newsletter even promotes Cook's "800" number, which has samples of new music and sound effects from his audio lab.

"One way we increase awareness about Cook Sound is through trade associations and conventions. We keep up a good network of potential clients that way," says Cook, who thinks that a face-to-face meeting with a potential client is the best sales tool of all.

What does the future hold for Cook Sound and Picture? More variety in audio. That's a certainty, because Cook says he believes in "niche- manship." Recently, Cook invested a quarter of a million dollars in a multi-track format, direct-to-disk (tapeless) system that makes production of audio tracks easier. "It was a tough decision," says Cook, who says the added flexibility of the digital audio mixes "will pay out well" with instant location of any part of a soundtrack, twice the quality of a CD, dialogue editing like a word processor, the ability to compose music on the

unit, the abili
and more.

"I will kee
my clients ne
have the pe
needs."

PART II
EQUIPMENT

4 Understanding Equipment and Needs

by Neil Heller

MATCH NEEDS TO EQUIPMENT

Production requirements determine the type of equipment you need. But the equipment purchasing decisions are never easy. One factor is usually how much money you have to spend. In addition, there is the knowledge that regardless of when you make a purchase, there is always something that will perform better and cost less just around the corner. The important factor is whether you can wait. There are many production and equipment purchasing decisions that cannot wait. To make the right decision, it is important to be knowledgeable about the capabilities and the limitations of available equipment in isolation and how pieces of equipment can fit into a component system.

GENERATIONS

There is an old saying that goes, "Everything is best at its beginning." This applies to video equipment in that the quality of a video signal is best at its beginning. To illustrate the point, think about the path of the video signal traveling through the various circuits from camera to switcher to videotape recorder in terms of viewing a picture and listening to a sound from the end of a tunnel. The longer the tunnel, the less clear the picture and the weaker the sound. Electrons are flowing through a wire. The longer the wire, the weaker the signal or quality of video.

That is why a video signal can be no better than the quality of video coming from the camera, or point of origin. Each piece of video equipment in the chain from the camera to the videotape recorder presents some form of opposition to the quality of the video signal. Manufacturers of equipment such as switchers, process and time base correctors attempt to build their product with one goal in mind: to make their product "transparent" to the video signal. Transparency means that the video quality at the input of a device is the same as the video quality at its output. Videotape recorders present the greatest challenge to video signals. However the amount of processing is so great and the amount of real estate available for recording is so limited that only with digital recording can real transparency be achieved.

The problem with most video recording is that the first recording is not used in the final production. Edited masters are usually the result of at least three generations of video recording. The first generation is created when the signal is recorded on tape, the second when the tape is edited and the third when the edited master is dubbed for final distribution. Although three generations are common, there are often more. When the format used for field acquisition is different from the editing system format, another generation is required to dub the field acquisition format to that of the editing system. Additional generations are also necessary to add video effects.

The difference between two commonly used

video terms for video production techniques is based on the number of generations each requires. ENG stands for Electronic News Gathering. Since the focus of news is on editorial content and not on the quality of the video image, news segments are usually the product of two or three generations of video. EFP, Electronic Field Production, is used for production where the emphasis is on both quality and content. The first generation video acquired for EFP must be of higher quality than for ENG as it will be subjected to more video processing and a greater number of tape generations in order to achieve its final product.

The number of generations required to create the final product is important because each generation of video recording decreases the quality of the video inputted by approximately fifty percent.

RESOLUTION AND SIGNAL-TO-NOISE RATIO

In a video camera or recorder, picture quality is determined largely by the resolution and signal-to-noise ratio specifications.

The greater each of these is in generation one, the greater each can be in the final product. The resolution specification for cameras is usually based on the number of horizontal lines a camera is capable of reproducing. The specification for signal to noise is the ratio of the amount of signal in the picture to the amount of noise.

Signal-to-noise is a ratio which compares the signal level to noise level in a video picture. The greater the intensity of noise, the poorer the quality of the picture. The measurement is expressed in decibels, with the signal noise set at a zero decibel reference. A signal-to noise ratio of 60db would represent an ideal video signal strength of one volt greater than picture noise. In reality camera signal-to-noise ratios of 54db and recorder signal-to-noise ratios of 45db are common figures.

Resolution is the amount and degree of detail in the video image. In general, resolution is determined by the amount and degree present in a specific bandwidth. The higher the resolving frequencies and the greater the bandwidth, the more details we can see in a picture. In standard video signals picture details are present in a band of frequencies from 0 to 4.2 MHZ. The color detail is contained in a 1.5 MHZ bandwidth centering around 3.58 HZ. The problem in composite sig-

nals is that room must be made for the color information, and this requires giving up some of the room allocated for luminance details. Color under systems express resolution in terms of the number of horizontal lines than can be recovered in playback. This figure is greater for black and white recordings than for color. Broadcast or component units express resolution in terms of a cut off point of 3db down at a frequency of greater than 4.2 MHZ. The higher this number, the greater the resolution.

COLOR UNDER RECORDING

As noted earlier, the goal of videotape recorder engineers has been to make the playback signals as transparent as possible to the recorder's input. This task has not been easy, and in many cases impossible. In general, videotape recorders can be placed in one of two categories. Those that record all the video frequencies directly on tape and those that must process video signals into other frequencies. The former system is usually reserved for high-quality videotape recorders. The latter has been used for recordings branded as "industrial" or "consumer grade." This indirect method of video recording takes into account smaller width videotape that lacks the real estate necessary to handle the higher frequencies required of direct recording systems. As a result, these frequencies have to be down-sized or converted. The 3.58 MHZ signal responsible for color information is "down converted" to a signal under 1 MHZ. The luminance or detail contained in a frequency bandwidth of 4.2 MHZ is converted to a bandwidth of between 1 to 1.6 MHZ. By placing the color information under that of the luminance these systems have become known as "color under," as the color information is positioned "under" that of the luminance. The advent of "color under" recording became the technological basis for the introduction of 3/4-inch recording systems in the early 1970s and 1/2-inch consumer formats with their industrial counterparts in the mid-1970s. The price of these video recording systems gave birth to the age of industrial recording.

This technology did not come about without some trade-offs in quality. Remember that two rules apply to a video signal and its representative frequencies: 1) Greater bandwidth will produce a higher resolution; 2) The higher the frequency the greater the detail. As a result the narrower bandwidths and lower frequencies of "color under" systems have a substantial effect on picture quality. The color resolution for these first generation video recording systems is only about 250 lines.

Low picture resolution is not the only problem facing "color under" systems. In fact, color and luminance separation is part of the video processing required for "color under" systems. The proximity of the color to the luminance signal means that sporadic frequencies radiating from both signals interfere with each other and result in the creation of unwanted artifacts.

3/4-inch SP, S-VHS, 8mm and Hi-8mm

It took almost 10 years from the introduction of 1/2-inch "color under" systems for video engineers to both widen the recording bandwidth and increase the recording frequency. These developments led to the introduction of the 3/4-inch SP format, S-VHS, 8mm and Hi-8mm. In reality, the technology that led to the development of these formats was not new. It had been present from the very beginnings of 3/4-inch recording. Signal processing in "color under" systems requires that the incoming video be broken down for recording into a signal containing color and detail. After the signal is recorded on tape, it is once again split into its color and luminance components for playback processing.

Y (luminance)/C(chrominance)

As a result, the concept of signal processing for "color under" systems is not new. Nevertheless, formats such as S-VHS and 3/4-inch SP still appear as second generation formats. Basically there are two reasons. The first concerns electronic circuits. Noise is a factor in high frequencies. At higher recording frequencies, a greater amount of noise is introduced into the system. Video engineers needed to design video processing circuits with highly accurate noise filtering systems capable of removing high-frequency noise with little loss in video detail.

The second factor is that recording higher frequencies requires that more tape real estate be made available. A specific section of tape is only capable of recording certain groups of frequencies. The smaller the tape the smaller the range of frequencies that can be recorded. The down sizing of videotape recorders in the early 1970s led to the development of the 3/4-inch recorder. The lack of

tape space required recording a narrower bandwidth and placing the color information under the luminance information. The result was the recording and reproduction of only 240 lines of resolution compared to greater than 330 for broadcast formats. From that point on the goal of videotape recorder designers has been to downsize tape formats and still maintain the highest possible quality.

COMPONENT RECORDERS

Since the greatest limitations involved the chroma or color information, one approach was to record separately the color and luminance information on different video tracks. This led to the development of the M and Betacam formats and later on M II and Betacam SP. Both the latter formats jointly record video in the form of Y, R–Y and B–Y components on 1/2-inch tape. Despite these similarities they are not compatible. They are commonly referred to as component recorders. Improvements in video technology also led to a second generation of recorders. For the defacto standard VHS this meant the introduction of the S-VHS and a source of additional confusion in the market.

Both S-VHS and Component formats base their improvements on the concept that separating the chroma and luminance elements of a video signal will result in a better video quality. However, only in the case of M II and Betacam SP can the video format truly be considered component. This is a result of the way in which the separate color and luminance signals are recorded on tape. In S-VHS both the chroma and luminance signals are recorded on a single video track as in the original 3/4-inch units.

SYSTEMS INTEGRATION

S-VHS, M II and Betacam SP have shown that video can be recorded and processed in many different ways. The key becomes integrating equipment into a system configuration. All the video equipment you purchase becomes part of a system. Video shot during field acquisition will become the source material for a post-production editing session. Shooting on one format, editing on another and distributing on a third has significant disadvantages.

For systems, the goal is to have the ability to use the acquisition video as the source for post-production editing. In general, "color under" systems can only maintain their quality over three generations, and even the quality of the third generation is somewhat questionable. Composite systems can go approximately five generations down. Acquiring your video in Betacam and having to dub it to 3/4-inch to use as a source will cost you one generation. This negates the quality gained in using the component recorder in the first place. The key is to be able to take the Betacam material and use it directly in a source Betacam unit. The ability to have a completely compatible system is usually more of a luxury than a reality because material is received from a variety of outside sources. Remember the following guidelines:

1. Invest your dollars in either the rental or purchase of the highest-quality video camera. As video is never better than at its source, the key element in determining production quality is the video camera.

2. If the final viewing destination of your production is conventional television monitors, the first generation recording on any format should be acceptable. The acquisition video format and source editing format should be the same. Avoid the duplication of tapes at all costs.

3. Master on a format of equal or greater quality. Master editing on a format of greater quality will lessen generation loss.

INTERFORMAT EDITING

Interformat editing is the ability to transfer information from the source of one format to the editing master recorder of another format. The capability to do this is the product of two features. Central to the systems is the editing controller with the capability to generate editing commands in the form of RS422 and the ability of both the source and the editing master to understand. RS422 machine control is independent of video formats and can be considered as a de facto standard of machine control. As long as all components of the editing system have this ability, it is not necessary that they be of the same format or even from the same manufacturer.

Interformat editing does bring up an additional consideration. If both formats are composite formats then a simple BNC single cable connection is all that is required to transfer video. However, if the difference is between composite and component video formats, then some form of transcoding will be required to take full advantage of the component recording. While this will yield better quality, it will also result in additional systems cost. Many manufacturers of component formats have recognized that the existing base of composite systems would be operational for years to come. Therefore, they have included composite inputs in addition to component inputs.

Taking the above information into consideration would yield one possible system configuration. Field acquisition can be done on a high-quality consumer or industrial/consumer version format such as Hi-8mm or S-VHS. The editing system can be configured with a like source unit, editing into a component recorder, such as an M II or a Betacam SP. The resulting master can then be used to dub copies in a distribution format such as VHS or 8mm. Although the dubbed copy will be three generations down, its quality will be considerably better than editing the same or lower-quality formats.

COMPUTER-GENERATED IMAGES

The signal quality advantages of recording in a component video mode do not carry over when computers become part of a video production system. Computers are being used with increasing frequency to create individual video frames for animation sequences. The direct output of computers is not in a form that is acceptable to standard video systems. Red, green, blue and sync cannot be fed directly into any conventional video recorder. While growing numbers of manufacturers are providing standard video interfacing, they do so at the cost of reduced resolution. As the use of computers in video applications grows, so does the realization that videotape recorders do not have the resolution required to reproduce computer-generated images. This is leading to the next generation of recording devices, the video disc. Not only do recordable videodisc systems offer the ability to record higher-resolution signals, a highly accurate ability to record individual frames makes them the perfect animation recorder.

Digital Systems

The age of computers is also accompanied by the increasing use of digital video effects devices. This market has witnessed a dramatic growth as capability has increased and prices have decreased. The key element in considering the use of both computers and digital effects units is the amount of processing contained in these units.

In digital video systems, the incoming signal is not consistent. Instead it is divided into sample sections that are assigned a number, 0 or 1. Once in digital form the signal can be manipulated in any number of ways. While resolution and signal-to-noise specifications hold the same importance as in analog processing, the more complex processing of converting analog to digital and digital to analog requires a new set of terms. The sampling rate of a video signal can be considered as the closest term to video resolution. It defines the number of times the analog signal is sampled and assigned a number within the period of a second. The higher the sampling rate the greater the resolution.

Sampling rates are always defined in terms of both luminance (Y) and chrominance (C) signals. Comparing the luminance to chrominance sets up a sampling ratio. The most common is 4:2:2. This specification tells us that the luminance bandwidth is double that of the chroma. This specification of double luminance to chroma bandwidth comes from our broadcast system. Older types of digital processing systems yielded signal ratios of 4:1:1 and lower resolution. Newer systems can digitally process video signals to yield chroma bandwidth ratios equal that of luminance. A typical specification would be 4:4:4. The combination of effects and paint systems has given rise to the adding of yet another four, for a 4:4:4:4 signal. The last four is actually a repeat of the first four. It stands for the luminance channel and it is mixed with colors to create a variety of intensity base effects.

Another important digital specification is quantization. This term tells us the length of the word the sampled point is converted into. The longer the word, the better that point of the video signal is defined. At present 8-bit quantization is the most common; however, future developments promise quantization rates of 10 and 16 bits.

When combined, both the sample and quantization rates result in the products output resolution. All digital processing equipment must con-vert the final output signal into analog for viewing and real-world processing. All signal processing from analog to digital and digital back to analog has the potential of producing a great deal of video noise. As a result, the frequency response of the analog has two critical responsibilities in digital processing equipment. First the input circuits must have enough bandwidth to appear to be transparent to the incoming video. Second the frequency response of the analog outputs circuits must match the sampling and quantization rates. All of this is expressed in the frequency response specification.

In considering the purchase of digital video equipment there are a number of things to keep in mind. Above all is the concept that digital processing can be a number of things and is not the end all solution.

- Digital equipment can be used to lock asynchronous signals together; this function is more commonly referred to as time base correction.
- Digital equipment can be used to store individual pictures; this is referred to as frame storage or frame synchronization.
- Digital equipment can be used to manipulate video inputs; this is digital video effects.
- Digital equipment can be used to created new images; this application is referred to as graphics generation or paint.

It is important to note that no attempt was made to define digital processing in terms of noise reduction. This is because digital processing actually adds a great deal of noise to the video. As a result this noise must be dealt with in either analog or digital noise reduction circuits.

SUMMARY: PUTTING THE SYSTEM TOGETHER

The system is the key element in the purchase of any piece of video equipment. Even the video person whose on-hand equipment assets are limited to a single camera or camcorder will have to consider the end result of his product in terms of post-production and distribution.

Any video system starts with the camera. A video can never be better than the quality at its origins. From that point any type of post- production work should adhere to the requirement that the format used for acquisition and the format of the editing source should be the same. Any other case will require creating a dub and result in the loss of a valuable generation of video. Likewise any digital processing equipment or digital effects used should have the ability of accepting a variety of formats. Component video recorders operating in composite mode produce a better signal than "color under" systems; however, the higher signal quality of working in the component mode is lost.

The grade format of the editing recorder should be at least as high as the source unit and preferably a higher-grade format. The latter situation will result in the least amount of signal loss.

Taking the system approach to video leads directly to the issue of purchasing or renting. While the financial considerations are best left to an accountant, the production needs rest totally on obtaining the best possible results. The bottom line is that no method of recording or digital processing can improve on the original video. The reverse is not exactly the same. Currently a number of high-quality consumer-type formats such as Hi-8mm and S-VHS will yield excellent first generation to third generation recordings. The key, once again, is the quality of the input video signal. Renting a high-quality three-chip camera to record on one of these formats will result in better quality than using a single-chip consumer format on a Betacam.

In this day and age when video technology is in a state of rapid flux, equipment rental could yield the best short term results and the best long term return on investment.

5 Purchasing Equipment: Three Case Studies

by Scott Carlberg

One lure of the video business is the image on the TV screen and all the ways to manipulate it. Another lure is the equipment—the awe inspiring feeling of walking into a control room that looks like the Starship Enterprise with track lights dimmed, a myriad of lights on racks of equipment, a steady electronic hum of activity and the frosty swish of airconditioning sweeping across your face.

That's precisely the reason purchasing video equipment can be a big problem. There is a desire to have it all, as if equipment defines success in video.

Success isn't measured in hardware. A fine video entrepreneur knows how to buy equipment that allows the company to meet the needs of its clients.

The case studies that follow have in common several traits of wise equipment purchases.

First, each company has defined the kind of customer it serves and the services it provides, which in turn define the equipment needs. No company tries to be all things to all people, thereby being nothing to any client.

Second, these companies don't gamble. Equipment decisions are not based on intuition. They are based on careful analysis of the client market and on what clients will pay to use.

Client analysis isn't enough, though. So, third, these companies look at more than the hardware itself. As Carl Levine, corporate television services manager of Unitel Video in New York says,

"We can't let marketing, engineering or any other department be the sole judge on the equipment we buy." These companies look at the way a new piece of equipment will affect them in every facet of business—the staff to support it, the clients who use it, the cost and the equipment's effect on productivity.

Any time a company spends $100,000, it's a significant event, no matter what is the size of the company. It's the kind of event that makes a manager stop and think twice. The companies in the following case studies have shown that they "think twice" in the planning stage of equipment purchases. They know their business strategies about the services they provide and the clients they serve—their framework for equipment purchasing. They understand that knowing the rules ahead of time makes their companies winners in the equipment purchasing game.

DRAWING ON EXPERIENCE: SPECTRUM VIDEO, CLEVELAND

For 10 years, Spectrum Productions has been a part of the Youngstown/Cleveland business community. It's an interformat production house with a client base that's 42% ad agency, 35% independent producers and 23% corporate work.

Spectrum has faced what many production companies have faced: video production has become much more than camera/recorder/editor. Clients ask for extra services, often in the form of "3-D" graphics and animation, requiring highly specialized and pricey pieces of hardware.

Spectrum saw its competition spending a lot of money on animation capabilities. There were several CubiComp units available at other Cleveland-area production houses. Spectrum made a commitment to 3-D in 1985, early in video animation development, with a system that delivered a good product for the time, but became outpaced by newer systems in competitors' shops.

It was time to either re-commit to video animation in order to meet the competition, or concentrate Spectrum's attentions on other video matters. "We determined that it was in our best interest to offer high-end 3D animation capabilities consistent with our other graphics and effects," says Doug Thorn, production manager for Spectrum.

With the decision made to compete in 3-D animation, how would Spectrum decide on a unit? Spectrum applies several thorough tests before making any equipment purchase.

First, does the proposed equipment fit the five year plan of the company? The five year plan is a creation of Spectrum's management team, made up of the department heads from engineering, operations, production, sales, graphics and the general manager. The plan doesn't always specify brands or models but does provide a focus for the kinds of clients and services of Spectrum. Says Thorn, "We look at where we're at and what we need to look at in the future to operate with quality. The difficult part of working out a capital budget is to include the non video items essential to the business—computers, office equipment and furniture—so we have to look at the whole operation and make the best decisions."

Whether the cost of the equipment fit in the capital structure, not just the facility structure, plays a big part in the decision, too. Spectrum paid cash for the $109,000 3-D unit. The company doesn't believe in financing equipment. "Financing costs can be burdensome." says Thorn.

Spectrum is part of WKBN-TV in Youngstown, Ohio, which gives it access to additional capital, but Spectrum's books are kept separate from the parent company, and they must be able to pay their own way with their cash reserve.

To be able to build that sort of cash reserve is either a luxury or the result of amazing financial

discipline. "It's the law," says Thorn, referring to the company policy what mandates that equipment will be paid in full, not financed. The parent company has a formula that applies to all Spectrum revenues, which builds a regular fund for capital purchases.

Second, how does the proposed equipment interface with the existing plant? "We have a basic equipment goal: to have equipment that will interface best with the money we have available," says Thorn, who notes that a company has to watch for hidden costs—changes that might be required in the existing system to make the new equipment interface easily.

Third, how does the market look? "Other people had the CubiComp. Cleveland-area clients have several places to go for that. Why should we be one more place to get it? We wanted to provide an alternative in the market." Customer reaction was more than speculation by Spectrum. The sales/marketing and graphics people interviewed clients in the market to project usage of an ultra-high-end 3-D animation unit at their company. Those interviews had to be positive to support buying the hardware.

Fourth, how long will it take to operate smoothly in our shop? "We look at the learning curve of the new unit," says Thorn. Spectrum does this by having the proposed units in its shop and checking how quickly his people learn the system. That's a good way to gauge the learning time, but doesn't give you an exact time. "Our learning curve was greater than I thought," says Thorn. "A high-end 3-D animation system requires programming skills, which most people don't automatically have. Also, the software is on the leading edge in the graphics world, so not even the manufacturers' trainees know all the answers when it comes to creating a specific illusion. It's a very technical and creative process, with emphasis on the technical."

Fifth, what's the track record of the unit and its manufacturer? Spectrum wants to avoid being a testing center for new hardware and checks with people who already have the equipment. "The unit we proposed had been tested a bit before we looked at it," says Thorn. "There were three (in Texas, Pittsburgh and Chicago), so we checked with those production houses."

After researching his market, Spectrum bought a Bosh Alias system for $109,000. Rental of the system from Spectrum is handled on a bid for each job, not on a "rate card." Thorn bases the cost for the jobs on:

1) the cost of the operator,
2) depreciated costs over five years,
3) a formula for factoring overhead and maintenance,
4) a margin for profit.

Six months later, Thorn's evaluation of the purchase was generally positive. "I think it fits with the level of service we wish to provide. It interfaces extremely well with our existing system and while we may have been overly optimistic in the training time required, our graphics people have produced some beautiful animation that our clients were very pleased with."

"IT'S NOT WHAT YOU MAKE, IT'S WHAT YOU SPEND": CREATIVE VIDEO, ATLANTA

Be flexible, don't add un-needed frills and watch how much you spend. That's the basic business philosophy of the co-owners—Jim Rocco and Guy Davidson of Creative Video in Atlanta.

When you look at the video business that way, some equipment decisions make themselves. Economics point out the hardware need.

Creative Video was started by Jim Rocco and Guy Davidson, who began their production house with a narrow focus: producing short promotional tapes for convention audiences for hotel cable systems in the Atlanta area. Creative Video was a basic camera/recorder operation that did its own editing in an outside facility.

After creating a successful production niche in the convention/hotel cable market, the "conventioneer" companies began asking Creative Video for specially produced corporate programs—not for hotel cable systems. Eventually, these non-hotel productions became Creative's major source of business, much more in line with standard video production facilities.

Creative Video's emphasis remained on producing—not post. In 1986, the year the company started, almost all of the company's equipment dollars were invested in hardware for "shooting." Post-production was done between 11 p.m. to 7 a.m. daily at someone else's suite. Rocco and Davidson had standing reservations with the edit company and paid a flat rate for the room.

As the business developed three editing bottle-

necks appeared. "Our time to post was too limited," says Guy Davidson, co-owner of Creative Video. "Eight hours a day wasn't enough for the amount of business we developed. Second, our clients were asking for fancier effects than we could deliver easily, and we had to go somewhere else even for basic effects." Third, the dollars spent for editing time, especially for the fancier effects, were rising rapidly and consistently, and Creative was paying off the note on someone else's equipment.

Davidson and Rocco were using a Grass Valley 100 switcher at the post house. They liked the 100, but needed visual effects beyond its abilities. They wanted access to more effects, the ability to layer certain effects and to save post time in their programs by eliminating "one effect at a time" editing.

The biggest impetus to making a change was their editing bills. "We were spending $10-11,000 a month on rented editing time—as much as the cost of buying a bigger switcher ourselves," says Davidson. It looked as though the bills would grow as business grew. Increased business would also make scheduling outside editing time a nightmare, and complications in editing time could hurt customer service.

"We are a market-driven company," says Rocco. "We serve our clients' needs. We had been a production company for four years and posted everything outside. When we looked at the numbers spent on edit time, it didn't take a rocket scientist to realize how we could save money—edit in-house." Creative Video decided to use the money they spent on edit time to pay off the loan to purchase equipment.

The team decided to build their own edit suite. Since the company already had video players and recorders, they didn't have to start from scratch. A switcher was the big-ticket item.

Most of the time in their edit suite would be devoted to posting programs shot by Creative Video. The rest of the time would be rented to other independent producers. In effect, Creative Video would become its own best editing customer, then outside producers would help pay the cost of buying the editing equipment.

Rocco and Davidson's goals for the edit suite were straightforward. It had to be flexible and reflect what their customers requested (an inter-format suite with high-quality basic equipment) with no unnecessary frills.

This tracks a key financial philosophy of Jim Rocco. "It's not how much you make, it's how much you spend that makes your company a success." Cost control is the key to financial viability, according to Rocco and spending that much without building equity in something ran against his grain.

That philosophy showed in choosing a switcher. The company wanted to stay with Grass Valley and looked at a Grass Valley 300, which had all the features they wanted but didn't suit the purpose of their edit suite. "We see the 300 as a switcher that's suited for broadcast instead of production work. It's fast and great for live work," says Rocco. A 300 also wasn't necessary to keep clients from heading to the competition, and Rocco says he doesn't try to win a hardware race as much as serve client needs. "It's easy to get into a 'I have more toys than you' mentality in this business," says Rocco.

How did Creative Video approach the equipment search? First, Davidson and Rocco stay current on video equipment. "The National Association of Broadcasters convention is the best way to learn about hardware," says Davidson, who thinks it's a good way to see all the equipment side by side. Second, "A person's background also plays a big part in the equipment decision process. For instance, I was raised on Grass Valley equipment, and you stick with the brands that you know and like," says Davidson.

Third, for the final decision, Davidson chose five manufacturers and explained to them the switching capabilities he wanted to have. He asked them to draw up proposals to meet those needs. Says Davidson, "The manufacturers came back with drawings and showed me the switching system they thought I'd need. I spent at least a day with each vendor to analyze and test their designs."

The final decision was in favor of the Grass Valley 200 switcher. "The 200 was a nice middle ground," says Davidson, who adds the 200 has the quality and the power to do the effects they want.

Davidson negotiated directly with the vendor on the unit he chose, and the manufacturer delivered the equipment through a local distributor. Creative Video handles its own engineering. However, Davidson used the local vendor as a system installer and as a source for purchasing ancillary equipment. "I used only one local vendor for other hardware and installation so I'd have a guarantee that the system would perform the way it should perform," says Davidson. Using one vendor avoids the problem of mixing equipment that isn't compatible because different vendors planned different specifications on equipment.

Creative Video could have opted for a cheaper switcher, but that might have compromised Rocco's main goal for the company—quality. "If you nickel-and-dime on quality to save money, pretty soon you'll be out of business," says Rocco, who notes that the quality has to show in order to sell clients. "You can't say to them, 'Come edit with me, I have a $25,000 routing system.'"

The concept of quality extends to Creative Video's engineering policy. Rocco doesn't skimp on test equipment to troubleshoot hardware problems. "The test equipment has to be fast. The time the equipment is down is time that revenue isn't being earned," he says. He has the same perspective on the engineering back-up from manufacturers or vendors—he wants quick response. Says Rocco, "I don't look at 10 hours of downtime in an edit suite as 10 hours. I look at it as $3,000 of lost revenue." Rocco's engineering back-up is consistent with his policy of quality and a strict preventative maintenance schedule.

When Creative Video bought the Grass Valley 200, it used a leasing agency's lease/purchase plan. Rocco and Davidson discussed the arrangement with several sales/leasing representatives. They wanted to compare interest rates as well as compare the reps' attitude about business. Rocco and Davidson are pleased with the arrangement they made. "The leasing company is easy to work with since it understands video, unlike so many financial institutions," says Rocco. "I always compare the leasing company to other financial institutions. This is how I make my choice about whom to work with. The leasing company gave me good advice, they know video, they are a good source of used equipment and they needed no money down. If I run into a problem, I know they don't want the equipment back; they'll help me with my problem."

The biggest disadvantage of a leasing company is that the interest rate is a fraction of a percent higher than what is available from a commercial lending institution. The other borrowing option, venture capital, is too costly for Rocco and Davidson. Rocco says the advantages of leasing companies outweigh the disadvantages.

Rocco believes that the key to success isn't how much you make but how much you spend, and his financing proves it's also "how you spend it."

GOING WITH "BIG RED": UNITEL, NEW YORK CITY

"When clients say you need it, you better listen," says Carl Levine, director of corporate television services for Unitel in New York City. That's Unitel's business philosophy in a sentence.

Unitel is a "facilities" company. It provides video equipment and the space to use it to people who have programs to produce. The company has three bases of operation: New York, Los Angeles and its mobile division based in Pittsburgh. The industry has counted on Unitel's services since 1969.

Any Unitel facility is a beehive of communications. "We have an active marketing staff receiving constant information from the markets we serve," says president John Hoffman in New York. "We listen and talk internally all the time." Peak internal communications is essential for a large company to keep every employee in tune with the latest business plans and to get the input of all employees involved with a business decision.

All Unitel employees are versed in corporate strategies.

First, Unitel takes a conservative approach to technology. "Fads don't translate into profit," says Carl Levine. "We want equipment and services with staying power." Therefore, Unitel tests the staying power of a service in the same way many video businesses do. Unitel often rents the hardware until the service proves that it will pay for itself, then it buys. To put the business strategy another way, according to Hoffman, "Remember, it was the pioneers of the old west who got shot in the back."

Second, Unitel knows how customers will use a piece of equipment before the company cuts a purchase order. "You have to have the work planned. It's all part of the budget," says Hoffman, who says a firm list of users is a primary part of the equipment justification process. "We don't buy equipment to woo people into our facility," he adds.

Clients come to Unitel with plans for series or special programs, and Unitel finds and/or adapts the right production space for the job. Before any equipment is bought or space is rented, Unitel knows the program, the client and the amount of money in the contract.

Third, Unitel emphasizes long-term contracts, and has focused on long-form programming for its clients. "There's little difference in the kinds of program in long form; it just depends on the story

the client has to tell," says Levine. Unitel has built its niche as a company known for creating the production facilities for continuing programs. Unitel's clients include MTV, IBM's field television network, Sally Jesse Raphael, Sesame Street and the Dow Jones/Wall Street Journal Report—a diverse group, all doing long-form programs. Unitel knows what business it's in; it has defined its client base and the kind of programming it will pursue.

It was with these business philosophies that Unitel faced a big equipment decision in November, 1987. The president of Unitel's mobile division, Dick Clouser, proposed a new tractor trailer be built and outfitted as a high-end production truck. (The truck has become known as "Big Red." All Unitel trucks are cream-colored, with the word, "Unitel" spelled in a bright color, which gives each truck its color name.) Price tag: $3 million. Clients: Top flight entertainment, such as classical and rock concerts, theater, broadcast specials and premier sports events. The proposal was accepted in 1987 and the truck was completed in April 1989.

Big Red was proposed at a time when there was no shortage of video production trucks in the United States. Some well-known names in video production were selling their trucks due to poor business conditions or because they were having a hard time making them pay. But Unitel has an advantage when it wants to attract customers—its name. Unitel is widely known as a high-quality, customer oriented company, which is the best kind of advertising advantage to have.

Unitel's experience in the mobile truck business gave it more than an edge in name recognition, too. Unitel is knowledgeable about remote trucks—who will use them and how to build them economically. "All we did was listen to our clients wishes and make an inventory of their wants," says Dick Clouser. "Then we developed the truck to fit their needs." Clouser points out that the only way to understand producers' needs is by working shoulder to shoulder with them. "You have to be out there with the people who use the equipment. Every day you spend away from the field is one more day away from understanding what it's really like in the field. You can only design a good truck by knowing field production."

Unitel applied its basic business philosophies to ensure the truck would follow the tradition of other Unitel facilities. "Fad" equipment wasn't part of the plan; high-quality basic hardware was. "Quality manufacturers—that's the key," says

Clouser, "because in this end of the business you need top of the line equipment. We're like a circus—we drive up and set up the tent for the show out in a cornfield somewhere—and there's no tech shop next store and no edit bay waiting close by if something breaks down. The equipment has to be dependable."

Clouser also follows another basic rule when trucks are designed and built—make equipment interchangeable. That doesn't mean that switchers or cameras must be the same model, but they must be compatible in order to add flexibility to the Unitel family of trucks. This interchangeability is part of the philosophy of total utilization.

Clouser also says that clients, not just technical quality, are reasons for choosing some hardware. "We choose manufacturers with accepted brands that our clients want. For instance, several years ago we had a truck with a certain brand of cameras that were great cameras, but didn't handle triax cable well. The cameras worked best in a fixed, studio configuration, which used heavier cables. Clients didn't want to use these cameras because of this lack of flexibility—not because of any picture problem. We had to change cameras to accommodate the wishes of the clients." That reinforces one of Clouser's tenets of equipment selection and purchasing. "It's not as big a gamble as it appears if you listen to your clients."

"High-end" doesn't just mean the quality of equipment, but how it's designed, too. Unitel says extra capabilities are part of "high-end" service, and Unitel tries to account for those extra's in every facility. "An example is a hockey game," says Clouser. "Any truck can do a four-camera regular season hockey game. But in the playoffs, any of our trucks, unlike most remote trucks, can cover the game and supply the same telecast to five or six feeds from the side of the truck. That's extra service."

Being ready with the extra capabilities is a foundation of Unitel's service philosophy. Says Clouser, "Having the extra capabilities, such as our high-performance audio boards in our trucks, may be considered overkill to some people. Not to me. It's like going to a party . . . if you don't know how to dress for the party, and you wear a tie and sport coat, you can always take off the tie to become more casual. If you go in casual clothes, you're out of luck if you need to be more dressed up." Unitel is "dressed up" for whatever may be requested in a shoot.

Unitel had contracts for the use of the Big Red before any equipment was ordered. Examples of programs for the new truck: Live from the Met, Professional baseball, top name prize fights and the Kennedy Center Christmas shows. At any given time Big Red has jobs booked five months ahead. While Unitel has built a high-end "niche" for its truck, it has guarded its income by varying the high-quality programs it handles—not limiting itself only, say, to sports or the theater.

The kind of jobs that Unitel books into Big Red help pay its cost, too. Where many trucks handle events occurring on the weekend, Unitel goes after longer production jobs, which may require four days to a week in one location, instead of two days as many normal sporting events can take. This policy keeps the truck in use for 72 % of the time, 365 days a year, increasing revenues for Big Red to $2 1/2 million annually, easily about a million dollars over other high-end remote trucks.

In effect, Unitel had the truck mostly paid for—they knew who their customers would be before the truck was ordered. Unitel worked with them every day, so there were no "cold calls" on customers to sell Big Red's services once it was ready to record. In a sense, the services were "pre-sold."

John Hoffman has certain other criteria that apply to all Unitel equipment purchases. Hardware is never considered in isolation but always analyzed in the context of the engineering and production people needed to make the hardware work properly. "We never make a unilateral decision on equipment," says Hoffman, who notes that the support people are part of the cost of running the equipment. As Clouser notes, "For a $125,000 camera, for instance, the person to maintain it and other cameras may run $90,000 with benefits. Retubing the camera may run another $65,000 for the life of the camera. All of that has to be considered."

Hoffman has different goals for equipment payout. He wants a three year payout on high-end gear and a five to seven year payout on basic video hardware. Equipment with brand new technology is more likely to be replaced sooner with newer technology, and Unitel wants to get the payout on hardware before it becomes "yesterday's news" and its revenue generating power diminishes.

Unitel applies its total utilization philosophy to every piece of hardware. For instance, a one-inch recorder may serve its time in an edit suite, then move to a mobile truck, then move to a duplication rack. Every piece of equipment is planned for multiple uses in the Unitel family.

PART III
BUSINESS AND FINANCE

6 Writing a Business Plan

by Neil Heller

The success of a business will be determined not only by how well the products and services are received but also by how well the business is operated. The requirements are the same, whether the business is part of a larger corporation or a smaller independent company. Whether you are a small private production company or a division of a larger non-video corporation, a clear understanding of your goals and how you will meet them is critical to your future existence. Starting a business of your own makes you an entrepreneur. But the spirit of independence is not limited to operating outside of a large corporate structure. Within the corporation the ability to make an independent stand is called "Intrapreneuring," a term coined by Gifford Pinchot III.

In either case, the key to accomplishing your goals rests in your ability to develop a solid business plan and have it work for you. A business plan starts with you and your ideas. It is a method to formalize your ideas, set them to paper and put them in motion.

The business plan starts with a title and definition of your business. The title should clearly define your goals and objectives. It should be short and to the point. The division is just as important as the name of a company. It gives both recognition and justification to the other parts of the company concerning the services you provide and the reason for your existence.

There are many ways of organizing a business plan. The following is only one recommendation.

THE EXECUTIVE SUMMARY

The executive summary is important. It is your speech to those who have, or you hope will have, an interest in your project. It defines what your company is, the reason why its services are needed and the previous experience of the principals involved in the company. The latter is important for giving justification for the whole concept.

For a new business, the executive summary should give a brief description regarding the physical location of the business and why you chose to locate in that area. It should also describe the active day to day role the principal(s) will be taking in the business. Finally, include any information on how the start-up of the company will be financed.

Although the executive summary comes first in a business plan, it is good practice to write it last. It is really a summary of the plan and should clearly set forth the concept and key elements of your business. A well-written summary is one that will spark the reader to finish the plan. Write your executive summary last and make several revisions until you are satisfied it does its job.

PART TWO

The next section should contain a brief description of each of the various components of the business.

Marketing Plan

The services you will provide, who you provide the services to, the geographical area your services cover and how you will promote your company are all essential components to cover in the marketing plan. You can elaborate further by adding a promotion plan which will explain how you intend to reach your customers and why they would elect to use you.

Competition

Competition is most easily defined in terms of the existing companies who provide similar services. Competition follows a marketing plan, as it is directly related to either a population group or a geographical area. For example, if you have defined your market as only the southeast, a similar business located in the northwest would not be considered your competition. You would only elaborate on the competition in your defined area. Once the competition is defined, you must justify the need for entering the field. Some examples would include providing a different type of service, one that a client will be enticed to use over the competition. Although you will be offering a similar service, list the components of your company that differ and improve upon the competition. Stating that you will offer the same service less expensively is not sufficient justification. In order to survive in the marketplace, you will have to be profitable. If you enter the field with the only advantage being lower cost to the client, what is to prevent your competition from doing the same and retaining old clients as well as attracting new, thereby destroying your client base?

The same considerations apply even in the corporate structure. Your division must be able to stand on its own two feet. Do not make the mistake of underestimating costs with the thought that the corporate overhead will support it.

Competition can also be expressed in simple supply and demand formulas. If demand exceeds the supply, and you can base your existence on meeting that overflow, that should be sufficient to justify your entering the marketplace.

Finally, competition plays a role even if your idea is new. Why haven't others picked up on this idea and what is the potential that competition will follow you.

Service

Define the specific services you will provide and how they will be provided. List the physical equipment needed and justify how that equipment will be used to provide your service. For example, if you are operating an independent production company, explain whether you offer in-house post-production services or specify that you will be using a third-party company. If you are operating a corporate media center, state whether you plan to rent your in-house post-production facilities to independents.

Staff

Determine the staff necessary to run your operation. Define the title and functions of each of your staff members. For in-house corporate centers, explain the effect of interaction between your staff and other members of the company. Discuss how your dealings with companies outside your own may affect your accounting department. Remember, if your plan calls for increased costs in other departments, they must be taken into account under your operating costs.

Vendors

Describe any special vendors or any particular group of vendors you will be dealing with. Explain the relationship you hope to have with them and what, if any, is your current relationship with these vendors.

PART THREE: FINANCIAL INFORMATION

This part is divided into two sections: capital and income. It is important to define how much capital is required to start the company, where it will come from and how much will be available to maintain operations once the company is started. This reserve is referred to as "capital reserves."

Income and expenses are next. Expect these numbers to change rapidly over the first year of operation. You will experience several "out of the ordinary" expenses that will occur during the first year in business. One thing is certain, expenses will be high and income will be low. This is normal operating procedure for a new business. For those members of the financial community the key will be your long term ability to stay in business and generate a profit.

As income and expenses will fluctuate during this period it is important to detail them in a monthly summary. Any trends or dramatic changes should be noted. Within the first couple of months you will be dealing with start-up costs. These costs should be detailed and an explanation offered as to what effect they will have on your first months of operation.

Next your income should be detailed. This usually means detailing your labor and supply costs. By determining your costs you will be able to calculate your required profit margins. This, in turn, will help you to set your pricing. Will you be able to sell your products and services at a competitive price? All this will lead to a return on investment (ROI). How long will it take you to recover your initial investment? How long will it take you to turn a profit?

Most overlooked in this area is a cash position statement. In many cases the billing to your clients will be 30 days. In today's business climate it is common to extend payment past the 30-day period. What will your cash position be during this period? On the payables side the same is true. Define your payment terms with your suppliers and how this will affect your cash position. Financial information, including forecasts and current statements, is essential.

CONCLUSION

As you can see all the parts of a business plan are related to one another. Working out a business plan is the best exercise for determining business success. Creativity and the ability to know the mechanics of your business is only one aspect. The other, and the most critical to your long-term success, is the business side.

By using this book as a planning tool, you can plan on success.

7 Understanding Business Accounting

by James Spalding, Jr.

First there was time, then barter, then money and finally accounting. Just as time keeps all things from happening at once, so does accounting keep all financial transactions from happening at once. Accounting is a way to measure your financial affairs at one moment in time and over time. Money is the measuring gauge of accounting and of your finances.

Accounting is necessary in order to comply with the minimum bookkeeping requirements of taxing authorities and, more important, to monitor the fiscal health of your business. You need to know which aspects of the business are doing well and which are doing poorly, which to expand and which to cut back. A good accounting system gives you the tools to keep the authorities happy and provides a feedback mechanism to a growing business.

Accounting tracks the natural business cycle: from young to old, in sickness and in health, a kind of micro-ecosystem. Like all systems, businesses have natural cycles. First some investment capital mixed with sweat equity, then initial expenses, accounts payable and cash outlays, next some sales, accounts receivable and cash receipts. Then we do some accounting to tally up the totals and see where we have been and how we stand. And luckily, we have enough left over to restart the cycle and reward the investors. And life goes on.

MONEY

Money is the measure of accounting, and money is nothing more than a medium of exchange. In the days of the ancients, sheep and horses were used to trade—three sheep and one horse for a healthy daughter. Some bright person, most commonly thought to have been the famous Phoenician, I. M. Money, figured out there had to be a better way to trade than with whole sheep and horses. And since cut up animal parts did not keep very well in the days before refrigeration, along came money as a means to facilitate barter and trade. Money itself does not have an intrinsic value, only the value that the traders attribute to it. Before the invention of paper money, gold and silver were commonly used. Since they have value as precious metals independent of coinage, it was easy for early traders to agree to their value as coins. But again, it was as a medium of exchange that money became important to society, and thus society valued money for what it could buy and not for its intrinsic value as a metal. Therefore, money, an underlying element of accounting, is an artificial medium with which to exchange sheep and horses.

Like money, accounting is a system of agreed upon contrivances. Accounting is an elaborate system that the modern business world uses by agreement in order to impose artificial order upon their financial affairs. However, like most other things, not everyone plays by the exact same rules, so the same transactions can be treated differently by different people. Accounting, therefore, is imposed order with very rational people disagreeing on some points.

That assets and liabilities are converted to a common denominator called money, with their flows being measured over time, is the cornerstone of accounting. But money and value are not constants, and time always passes. There are inflation, national deficits, and international exchange rates. These are all indicators of the ebbs and flows of the financial world. Is the house worth the $300,000 you paid for it, or the $400,000 someone is willing to pay you for it now? The next contrivance is to allow the past to stand. Historical values are used in accounting to measure what happened when it did. You really paid $300,000 and your mortgage is $210,000. Using historical cost works fairly well with short-lived assets and low inflation rates, and not so well with long-lived assets and high inflation rates. Historical cost is the basic contrivance for determining at what value a transaction is measured. It is what happened, not what could have or might have been. However, a savvy manager will not forget what the market value is regardless of the rules of accounting.

Accounting is not an intuitive art or an empirical science. At the back of this chapter are an overview of an accounting system and detail examples of accounting schedules and financial statements. The "Facility Business Company" used is fictitious, but the formats and income and expense types and categories are taken from real facility businesses. These will be referenced in the chapter. The examples are meant to bring a dash of reality to these abstract accounting discussions.

FINANCIAL STATEMENTS

Accounting reports that give a picture of the financial affairs of a company at a given moment in time are called balance sheets. Those that show a picture over time are called income or profit and loss statements. The balance sheet deals with assets, liabilities and owner's equity. Assets are the things that are owed to you or owned by you, such as cash, accounts receivable and equipment. Liabilities are what you owe to others, such as trade payables, notes payable and other forms of debt. Equities are the excess of assets over your liabilities, that is, the excess of what you own or are owed over what you owe. For example, you paid $300,000 for your house and owe the bank $210,000 on the mortgage. Here the house is your asset at $300,000 with liabilities at $210,000, with your net worth or equity in the house at $90,000. After the next mortgage payment, your liabilities are lower and your equities higher.

The income statement deals with the use of assets and liabilities over time. Income is receipts or earnings for services or delivery of product. Expenses are expenditures for services and for non-capital assets and for depreciation on capital assets used to produce income or maintain the business. The net profit or loss is the excess or shortfall of income compared with expenses. If income is higher, there are profits; if expenses are bigger than income, a loss results.

Figure 7.1, Financial Statement Grid, shows accounting's big picture—how balance sheets and profit and loss statements (P & L's) are formatted. The two sides of the gird are related. Profits increase owner's equity or net worth, and losses decrease equity. For an example of facilities type financial statements, see Example 5.

Debits and Credits

Our next adventure is to the world of debits (DR) and credits (CR). These are definitional increases and decreases on the balance sheets and profit and loss statement. They are accounting conventions, and it is advantageous to memorize the definitions. A banker's credit is your debit and vice versa. A bank's books are a mirror image of your account and need to be transposed to work for you. The grid in Figure 7.2 shows how debits and credits affect the financials.

By definition debits are increases in assets and expenses and decreases in liabilities, equities and income. Credits are increases in liabilities, equities and income and decreases in assets and expenses.

Double Entry: Debits Equal Credits

Because all things are not created equal, the accountants of the Middle Ages created a system where all entries must balance out, with all debits equaling all credits. This was called double entry bookkeeping. The debits were on the left, and the credits were on the right of two-column ledger paper. (See Table 7.1.)

These merchants obtained $1,000 in spice and had incurred a $500 dockage fee, for a total cost outlay of $1,500 in gold. Note the double entry— one side for debits and one side for credits. More than one entry per side is permitted. The next trick is to have all the debits be in balance with the credit side. Doing it correctly includes understanding each transaction and translating it to a

Figure 7.1: Financial Statement Grid

BALANCE SHEET (a company's financial picture at a given moment)	INCOME STATEMENT (picture over time)
Plus (+) ASSETS	Plus (+) INCOME
Owned by or owed to the business—things having value to the business.	Receipts and earnings for services rendered and product delivered.
Minus (-) LIABILITIES	Minus (-) EXPENSES
Debts owed to creditors for services and goods received. Monies borrowed.	Services and goods used to create income or maintain the business.
= EQUITY or NET WORTH	= PROFIT OR LOSS
Assets less liabilities equal the net worth or the owner's equity in the business.	Income less expenses equal net profit, if positive, or net loss, if negative.

Figure 7.2: Debit and Credit Grid

+ ASSETS	+ INCOME
DR = Increase	DR = Decrease
CR = Decrease	CR = Increase
− LIABILITIES	− EXPENSE
DR = Decrease	DR = Increase
CR = Increase	CR = Decrease
= EQUITIES & NET WORTH	= PROFIT and LOSS
DR = Decrease	DR = Decrease in profits or an increase in losses
CR = Increase	CR = Increase in profits or a decrease in losses

Table 7.1: Double Entry Bookkeeping Sample

	DR	CR
Spice inventory	$1,000	
Dockage fee	500	
Gold expended		$1,500
Totals	$1,500	$1,500

bookkeeping entry. No one can pretend that debits (DR) and credits (CR) is easy, but with practice . . . it becomes easier.

The Spread Sheet

All the double entries for each transaction get collated into financial statement summaries. The usual method is to collate like-kind transactions by using a spread sheet, either electronic or manual. For instance, all cash paid out is summarized in the **cash disbursements journal.** One column is for cash out and the others categorize the type of expenditure. (See Table 7.2.)

Note that the entries are in balance. The total of the cash out column (credits) is equal to the sum of all the types of expenditures (Payroll, Office, Postage & Miscellaneous) (debits). A similar journal is prepared for the cash receipts and other significant groups of transactions. These journals (see Example 2 - Cash Receipts and Disbursement Journal), which summarize the individual transactions, are themselves summarized by the Trial Balance, a specialized spreadsheet that lists all assets, liabilities, equities, and income and expenses (see Example 3 - Trial Balance). Remember that debits must equal credits. Position these summarized numbers on the trial balance that is arranged in a fashion similar to the financial statement grid. Then balance debits and credits to insure accurate transcription. Thus raw data for the financial statements are tabulated.

Table 7.2: Cash Disbursements Journal Sample

Paid to:		Cash	Payroll	Office	Postage	Misc
145	Handy Inc.	$100		80	20	
146	J. Help	500	500			
147	Ms. Post	300		100	150	50
Totals		$900	500	180	170	50

ART AND ARTIFICE

The next step is where the juggling and artwork comes in. An accounting system groups like-kind transactions together into a financial statement format. The key to formatting meaningful financial statements is to track meaningful information. Meaningful financial information provides reports that monitor the key economic variables of a business through and over time. Tracking these key economic variables of income and expense can provide the feedback for success. In addition, the non-critical variables need to be reviewed to insure that they are still non-critical. What is critical to a business varies over time, so the accounting system needs some flexibility. Early on in a business' life, sales and cash may be critical and must be monitored closely. As the business matures and the comfort levels accumulate overhead may need closer scrutiny.

In designing an information system, it is important to know what will be retained and for how long. Whatever is coded into a system and retained as a basic building block can be grouped and regrouped. If you have already summarized and deleted a specific detail, how you have it summarized is how you have it—unless you want to re-enter the detail, if it is available.

For example, if you summarize at the income and expense level—with no details, you would not know what types of income there were and what types of expenses were incurred. There would be just two lump numbers called income and expense. A more meaningful way would be to categorize income by client: corporate, advertising, broadcast/ entertainment and independent producer; or by service: duplication, video editing, graphics and audio. Then within each of these major types develop subtypes by key revenue centers such as edit suites, specialty machines and services.

The general income statement format is as follows (see Example 5: Financial Statement, in Figure 7.8 for details):

Income—Cash received or revenues earned are tracked by type with the largest volume source first, the second largest next and so on. The "type" can be difficult to pick and you may want to allow your sales files data managers, the information buckets who collect sales data, to report sales by differing criteria. For example, you may want to report all sales by equipment used [edit suite #6 with its subcomponents (tape machines used, online edit fees, special effects and so on)], or you may want to have the sales categories be client driven such as broadcast, advertising, production companies, corporate and miscellaneous clients.

If your sale data files are flexible enough, both client and service tracking can occur so both types of economic variables can be monitored.

Cost of Sales—This grouping is for variable

Figure 7.3: Income Statement Format

+ Income—Cash Received or Revenues Earned
− Cost of Sales—Costs of Product and Services Sold
= Gross Profits—Income less Cost of Sales
− Overhead—General and Administrative Expenses
= Net Profits—Gross Profits less Overhead

Table 7.3: Income Types

Client Driven	Service Driven
• Advertising	• Production
• Corporate	• Editing
• Broadcast/Entertainment	• Graphics
• Motion Pictures	• Duplication

expenses that are incurred to create sales dollars. In a business that sells product, cost of sales is fairly easy to define—it is the cost of the product (called inventory) that is sold and some incidental expenses such as delivery cost and commissions. Raw tape stock sold to the client direct is an inventory cost of sales. For a service company, like most facilities, the lines get a little blurred. Essentially, variable costs are those expenses that would not have occurred except for a particular sale. Overhead expenses, those that are incurred as a cost of being in business, are in contrast to variable expenses. There can be lots of gray areas. Decide and sort; review each year. For example, when you add a new system and bring in freelance operators as clients demand, the operator is a variable expense of the sale. Later, after you have departmentalized the company, the entire post production department can be treated as a cost of sales. The goal is to break out variable expenses from fixed expenses.

The next layer is to format the cost of sales expenses parallel with the income groupings. It is important to relate the variable cost of sales to the sales groupings that caused the expense to be incurred. The goal is to track sales and expenses by type.

Gross Profit—This is income less variable cost of sales. If you set the two of them up so you can track sales, variable expenses and cost of sales, you can tell if you are making any money before overhead. Of course, you should already know whether you are covering costs, but by comparing different gross profit types you can see which ones contribute the most in absolute terms. And by making a percentage of sales comparison, you see which sales types are best per sales dollar. By knowing where the money is coming from, segments of the business can be nurtured and other segments cut back.

Direct Expenses—As companies grow and departmentalize, an intermediate cost grouping oc-

curs. Sales subgroupings, usually along functional lines, such as production, postproduction, audio and special effects will have administrative support that is for the particular segment of the business. It is on-going and not directly caused by particular sales, but is not a company-wide effort. Scheduling, engineering, and quality control and supervision are direct expenses of certain sales segments, but not general to the business. These are considered direct expenses of the sales segments and departmental overhead and not general overhead. When analyzing profitability, these direct expenses need to be considered with the appropriate sales segments.

Overhead—General and Administrative Expenses—These are fixed costs, given the present operations of the business. These would include rent, utilities, basic telephone charges, postage, office supplies, and other similar necessary expenses that are not directly related to sales. Also included with overhead-type expenses are those expenses that are impractical to allocate to cost of sales—such as long-distance charges, copy costs, operating supplies, and others. Overhead expenses need to be monitored to see that they are reasonable in relation to sales volumes and prior experience.

Consistency—This means it is important to compare apples to apples and oranges to oranges. Be consistent in categorizing income and expenses. One of the best ways to monitor how a business is doing is to compare it to itself. This can be done by comparing the current month to the year to date totals, and comparing year-to-date totals to prior years. This is only meaningful if the underlying data is consistently grouped the same way.

CASH VERSUS ACCRUAL ACCOUNTING

There are only a few accounting jokes, so listen up. Everyone knows the answer to two plus two. But for those of you who don't know, the answer is four. However, for the erudite university mathematics professor type, the answer is not so simple. For example, if you are dealing with a base system of three, two plus two is equal to 11, and that is assuming real numbers. Imaginary numbers raise many other concerns. And you thought the answer was easy! Now ask your friendly Certified Public Accountant, commonly known as a CPA. If the CPA is worth his salt, he'll ask "what does the client want it to be?" It's meant to be funny, but there is more than a grain of truth in the story. Different methods of accounting produce different answers. And, assuming mathematical and system integrity, both answers are correct. The key to this apparent contradiction is that, by selecting different accounting methods, you have asked different questions.

The two accounting methods most widely used by professionals and businesses are the **cash method,** more formally known as the cash receipts and disbursements method and the **accrual method. Cash flow** is not a method but a useful planning tool. Cash flow management will be looked at later. Finally, the fund accounting method is used by governments and institutions to track restricted funding and special purpose expenditures. Since fund accounting is very limited in its use, no further discussion here is warranted.

Cash Method

The cash receipts and disbursement method is by far the method more commonly used by sole proprietors and small businesses. Its principal virtue is its simplicity. Revenues are recorded when received: suppliers' bills get recorded when paid. For most small businesses this works well. This usually works well for taxes also since income is not taxable until received, and expenses are not deductible until paid. Sounds fair enough. And it usually works in uncomplicated situations where the timing of receipts and checks disbursed are fairly close together. It's simple and it usually produces reasonable information. This is similar to checkbook accounting most people use for their personal finances. But there are some complications. Assets having a multi-year life and prepayments (insurance, interest, rental deposits and so on) must be written off over their useful lives, or when earned by external entities.

Accrual Method

The foundation of **accrual accounting** is the concept of matching revenues and expenses in the same accounting period they are earned and used. Income and expenses are determined regardless of when the check is in the mail. Its cornerstone is the concept of an **enforceable claim.** Revenue and expenses appear on the income statement when the supplier earns the income and, if necessary, could go to court and legally collect it and when the customer receives the goods and services and is legally obligated to pay the supplier. Revenue or income is recorded in the period it is earned, regardless of when paid. Expenses are recorded in the period they are incurred, regardless of when paid. The typical time lag between revenue being earned and the money being received creates **accounts receivable.** The time difference between expenses being incurred and the bills being paid creates **accounts payable.** Later, when checks are exchanged, no revenues or expenses are recorded, just the reduction of accounts receivable and payable.

An Example Using Both Methods

We'll use Now You See It Inc. and its stellar employee, I. M. Doolittle. The company pays its employees monthly three days after the close of the month. Mr. Doolittle has worked every work day during the month at the rate of $2,000 per month. At the end of the month he has a legally enforceable claim of $2,000 (pre-tax) from Now You See It Inc. Mr. Doolittle is on the cash basis of accounting and the company is on the accrual basis. Mr. Doolittle does not record his pay check as income until he receives it on the third day of the month following. That's the cash method— record a transaction when cash changes hands. The company however, per the accrual basis, records the expense at the month ending with an offset to payroll payable. On the third following, the company pays Mr. Doolittle $2,000 gross (he actually receives his net pay after taxes). When Mr. Doolittle is paid, the company reduces its payroll payable with offsets to cash and the employee withholding taxes due the government. The cash basis shows the payroll when paid. The accrual basis shows the same payroll on the company's books when earned by the employee, when an enforceable claim exists.

Guidelines for Accrual Accounting

Income—When the income is earned is when you book the revenue regardless of when paid. This assumes, however, that it does get paid.

Expenses—Salaries and wages are booked at the time the labor is performed.

—Consulting and other services are recorded in the period the services are rendered.

—Goods and services are booked in the period they are delivered or received.

When to Use Which

Cash basis accounting is preferable for the smaller business where the owners can keep track of timing differences between sales and expenses. It is also usually beneficial for tax purposes, if accounts receivable are larger than accounts payable. If a business pays its bills faster than it receives monies for its invoices, using cash basis accounting avoids paying taxes on receivables until the cash is in hand.

Accrual accounting makes sense when a business receives or pays for financial transactions in periods different from when they occur. For example, assume a business has received a prepayment of $10,000 in May for work scheduled to be performed in July. Under an accrual system, the income is deferred until the work is performed in July. Accrual accounting is usually preferable when a business has more than one person in charge of different jobs, often the $300,000 to $500,000 range of annual sales volume.

A business can keep its books on the cash basis for taxes and on the accrual basis for internal reporting and job analysis. This allows for good information systems and often minimizes taxes. All this may seem a bit complex for a techie, artist or even a manager. But there are bookkeepers, accountants and CPA types who can help you get set up and help deal with your accounting, financial, and tax questions, and assist as needed. Although you don't need to be a rocket scientist to deal with financial affairs, a good professional can help with the more erudite matters and steer you clear of the many pitfalls of the financial world.

CASH FLOW MANAGEMENT

Clients are wonderful, high monthly billings are great, but cash is the *sine que non* of business survival. You can't pay your rent or feed the kids with accounts receivable. Cash in the bank and in your pocket can take care of your needs, nurture your business and allow you to sleep at night. Obviously a business doesn't start out with much cash in the bank—it takes happy clients and then billings to generate operating cash flow. You can, however, arrange your business affairs to speed up the conversion of your services and products to your clients into cash in your hands.

Cash Flow

Cash flow is a commonly used financial term. Taken literally, it means cash flow-in and cash flow-out over a specific period of time, with the difference being the net change in cash balances. For example, cash receipts for the month are $10,000 and $8,000 has been paid out in checks and petty cash. The cash flow-in is $10,000, the cash flow-out is $8,000 and the net cash flow is a positive $2,000, an increase in cash balances. As a financial tool, the cash flow concept is used to project future cash-in and cash-out to determine whether there will be cash surpluses or shortages. This is another spread sheet application. It is usually done on a monthly, quarterly or annual basis, using history, trends and astute guesses about the future to make the projections. The results can be positive or negative cash flows. Too much cash calls for investment decisions. Too little cash means taking corrective actions to improve cash flow, or, if it is a temporary shortfall, see about getting a loan.

Cash flow is not an accurate indicator of whether a business is profitable. You can have a positive cash flow while incurring business losses just be selling off assets or by not paying bills. Or you can be profitable with negative cash flow due to slow paying clients. Cash flow is not an operating accounting system unless used with the cash basis method of accounting. Cash flow is generally used to project the future cash requirements of the business. It is a financial tool, not a system.

Cash flow management is using business techniques to improve how you deal with cash. The **cash receipts** side is grouped into terms and conditions of sale, prompt billings, accounts receivable agings, collections and deadbeats. The **cash disbursements** side is more limited, but with more control—here we look at cash discounts, timing of vendor payments and prioritizing your payables.

Cash Flow In

Terms and Conditions of Sale—The trade terms of a business are the stated provisions of the sales agreement with a client. To be enforceable, they need to be in writing and signed by the client. The easiest way to do this to have them printed on the sales order (usually the back), with the sales order being signed by the client, e.g., "I agree to purchase the above items at the specified prices and have read and agreed to the terms and conditions on the back of this sales order.'' Trade terms can get complicated, but they should include prices or a reference to a prevailing rate card, delivery terms (who is responsible for shipping and insurance), methods of acceptance, and payment terms. It is primarily the payment terms that affect cash flow. For large jobs you can request production advances and progress payments, so you can use the client's money to finance their job. It won't work all the time, but you will never get an advance payment if you don't ask.

The other part of payment terms is how fast payments are due upon invoicing. Do you ask for the net balance within 30 days of invoicing ("net 30 days"), or should you offer a cash discount—a 2% discount if payment is received within 10 days of invoice date ("2%/10, net 30")? Cash discounts are a direct tradeoff between discount taken expense, the 2% of sales, versus the interest that could be earned by the faster payments on some of the accounts receivable. The break-even point occurs when the imputed interest earned on the decrease in receivables equals the discount taken expense. For example, the average sales are $10,000 per month, and the average number of days receivables are outstanding is 60 days. Next you offer a 2%/10 discount—2% off the sales price if the invoice is paid within 10 days. Twenty-five percent of your clients take advantage of this discount so your days receivables outstanding drop to an average of 30 days. Push the numbers through.

$10,000 \times 25\% \times 2\% = \$50 =$ discount taken.

Average days receivable outstanding goes from 60 to 30 days and means a speed up of one month's sales receipts. Assume 12% interest, or 1% per month.

$10,000 \times 1\% = \$100 =$ the imputed interest that could be earned.

The discount expense is $50; the imputed interest is $100.

The facts have been skewed for simplicity; they could be rearranged to have the discount expense be greater than the imputed interest. However, the exercise does show the important steps in deciding whether to offer a cash discount. Other factors to consider are whether a cash discount will increase your sales volume, and possibly improve your bad debt losses, both likely to be positive factors but hard to quantify. Your payment terms—cash up front, cash discount to cash whenever—do affect your cash flow. The looser

the terms, the higher the sales, the slower the payments, the higher the bad debt. Tighter terms means lower sales, faster payment and more certainty of payment.

Prompt Billing   This is a critical step in cash flow management. The faster you bill, the faster you will get paid. If a bill needs multiple approvals before being sent out, set up standard operating procedures to streamline the process. If specific knowledge is the bottleneck, invest time and personnel in educating a skilled billing coordinator. It is often easy to get bills out fast in a post-production fax mill where quantity used times standard rates are givens. In the multi-task and staffed production world, efforts are often rewarded by requiring and standardizing prompt personnel and equipment usage time reports. The more standardized the job units, the easier the billing process. The more custom or specialized the job, the more decision time in the billing process.

Another frequent billing bottleneck is holding up invoices for minor items such as the exact cost of shipping or duplication. There are two ways to handle the bill delayers. First, you can estimate, either from prior experience or a phone call to the vendor, the amount within a reasonable latitude. To the penny, and often to the dollar, exactitude is generally not required if honest and unbiased efforts are made to estimate minor amounts. The second method is to bill for all known amounts, and have a follow-up bill to catch the late arrivals. State on the initial bill that a follow-up will be forthcoming. The second method can usually be used where the margin of error in estimating is unacceptable to you or to your client. Either method is preferable to holding up 80% to 90% of an invoice over minor amounts. Get the billing out as fast as possible.

Accounts Receivable Aging   Aging your accounts receivable is not a Ponce de Leon trick. It is spreading each and all receivables by period of occurrence by client. (See Table 7.4.)

Each line is a summary of a client's activity by month. Each column is totalled at the end of the client name list. This aging schedule is usually prepared on a monthly basis. The focus on the receivables' aging is large account balances regardless of age, and account balances that are growing older, largest amounts being the most important. This listing gives you a visual aid to each client's activity and how fast each pays. Non-current accounts require some collection efforts.

Table 7.4: Aging Schedule Sample

Client Name:	Total Due	Current Month	30–60 Days Prior	60–90 Days	90 Days and Over
Each	$5,000	$1,000	$3,000		$1,000
Client	$4,000	2,000	1,000	$1,000	
Total	$9,000	$3,000	$4,000	$1000	
Percent Total	100 %	33 %	45 %	11 %	11 %

The accounts receivable aging schedule is an important tool to allow you to project the next month's cash receipts and help you focus your collection efforts for their maximum results.

Collection   In two words—**follow-up, follow-up.** People may want your services and products, but they may not always pay you, especially when you want to be paid. The first phone call—they never received the bill. A copy goes out in tonight's mail. The next call discovers the bill is wrong. Talk to your sales and production people to find out what was actually promised and then actually delivered. Why wasn't that on the paperwork? Adjusted invoice goes out. Now you can begin asking about the payment. Most collection work is not harassing sleazo clients, but rather being a catalyst for clearing up billing problems. And when the client is a big bureaucracy, you may have to push your invoice through every level of approval and payment, and get paid late anyway. For repeat clients, try to negotiate faster payments under the terms and conditions of the sales agreement. Even bureaucracies have ways of paying bills fast when they want to.

So collection efforts need to be frequent, friendly and positive—you don't want your collection efforts to offend a valuable client—and persistent. Keep calling. The squeaky wheel does get oiled!

Deadbeats   There are slow paying valuable clients and then there are deadbeats. Prevention is the best cure. While no cure is 100%, reasonable credit applications with references checked before the sale, and a sharing of slow payer information with your competition, can help minimize deadbeats even coming through your doors. Let's assume it's too late for the cure. Each deadbeat needs to be evaluated with final collection efforts following. If the firm is going under, try to work out the best deal you can—consider goods in lieu of money. Be sure to fill out all the bankruptcy forms. That happens and is a normal business risk. For the other deadbeats that need your money pried from them, start out with a letter on your stationery, followed up by a letter from your lawyers, then court. For small amounts, usually under $5,000, represent yourself at small claims court. Remember to keep collection costs in mind while pursuing deadbeats; if costs exceed the original bill, you have gone backwards. Generally a tough attitude towards deadbeats, tempered with financial reason, will work in the long run.

Cash Flow Out

Cash going out the door is much more controllable than cash flow in—you decide when to write the check. The only time this is not the case is when your cash flow is negligible and your suppliers want their money.

Cash Discounts   Cash for the purchaser represents the same weighing of cost versus benefit that was discussed with trade terms. If you can save more than it costs you to borrow the money, go for it. For example, you can get a 2% discount in the purchase price by paying a little faster. If you assume you have to borrow the money to pay faster, it's a straight comparison of interest costs versus discounts earned. The window to look at is the discount earned versus the interest carrying cost to the date the bill would normally be paid. Assume you normally pay in 30 days and an interest rate of 12% per year or 1% per month. By paying early you can earn $2 per $100 of purchases.

The cost would be 1% for two/thirds of a month (20 days): $1\% \times \frac{2}{3} \times \$100 = \$.67$.

In this example, you earn $2 through discounts taken, and spend $.67 in finance charges. Given these assumptions, it is wise to take the discount. And that is usually the answer when discounts are

Figure 7.8: Example 1: Overview of an Accounting System

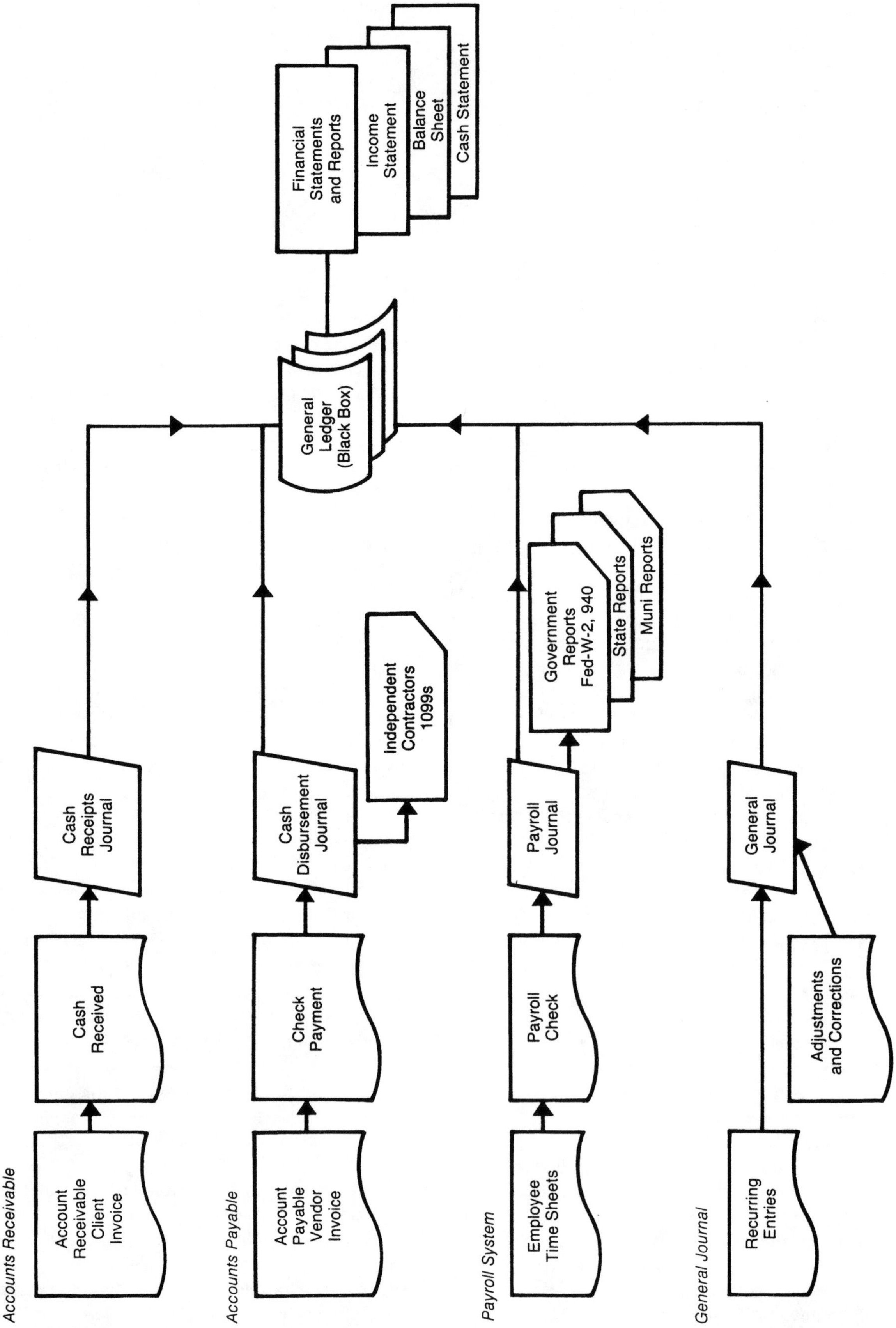

Figure 7.8: Example 2: Cash Receipts and Disbursement Journals

Facility Business Company — Cash Receipts and Disbursements Journals' Summaries — 2/15/96

1995	Jan	Feb	Mar	Apr	May	Jun	Jul	Aug	Sep	Oct	Nov	1995 Dec	1995 Y.T.D. 12 Mos	1994 12 Mos
RECEIPTS														
Postproduction														
Editing	25439	35189	34653	40511	45333	35915	33263	46932	35664	48963	49984	40331	$472,077	$420,189
Graphics	9226	13966	10985	15981	18935	13556	12890	18902	10315	15669	16905	15802	173,132	162,231
Audio	7543	5001	3006	10890	6455	12469	12999	10553	3003	5556	15680	5559	98,714	89,429
Other	1321	140	1009	2589	1159	4006	5566	1008	1111	2645	2369	5226	28,149	27,895
Postproduction	43529	54296	49653	69971	71882	65946	64718	77395	50093	72833	84838	66918	$772,072	$699,744
Duplication	1896	2001	2260	1533	1389	5559	1633	1222	895	1667	1222	3522	24,799	21,215
Production	8663	14663	16936	19654	11236	11695	22963	11445	12396	18963	11569	11300	171,483	170,459
Interest	75	59	40	33	102	306	298	313	236	132	93	80	1,767	2,170
Other		527			2000			456			56		3,039	6,009
Total Receipts	54163	71546	68889	91191	86609	83506	89612	90831	63620	93595	97778	81820	$973,160	$899,597
DISBURSEMENTS	5.57%	7.35%	7.08%	9.37%	8.90%	8.58%	9.21%	9.33%	6.54%	9.62%	10.05%	8.41%	100.00%	108.18%
Cost of Sales:														
Talent/Script	1119	1479	1424	1885	1790	1726	1852	1877	1315	1934	2021	1691	$20,112	$18,945
Videotape	4538	5994	5772	7640	7256	6996	7508	7610	5330	7841	8192	6855	81,532	75,369
Production Expense	3511	4638	4466	5912	5615	5414	5810	5889	4124	6068	6339	5304	63,090	58,321
Freight	1260	1664	1602	2121	2014	1942	2084	2113	1480	2177	2274	1903	22,634	21,865
Travel	1966	2597	2501	3311	3144	3032	3253	3298	2310	3398	3550	2970	35,330	32,659
Diem	1092	1443	1389	1839	1746	1684	1807	1831	1283	1887	1971	1650	19,261	16,899
Set Costs	388	512	493	653	620	598	642	650	456	670	700	586	6,969	6,490
Location Fees	220	290	279	370	351	339	363	368	258	379	396	332	3,945	3,599
Outside Services	5368	7091	6827	9037	8583	8276	8881	9002	6305	9276	9690	8109	96,445	85,643
Film Costs	271	358	345	456	433	418	448	454	318	468	489	409	4,868	4,500
Total COS	19733	26066	25098	33224	31552	30425	32648	33092	23179	34098	35622	29809	$354,546	$324,290
General Expenses														
Salaries & Wages	14772	19513	18788	24871	23621	22775	24440	24773	17351	25526	26667	22315	$265,412	$241,847
Payroll Taxes	1930	2549	2454	3249	3085	2975	3192	3236	2266	3334	3483	2915	34,668	31,198
Employee Benefit	757	1000	962	1274	1210	1167	1252	1269	889	1308	1366	1143	13,597	12,092
Legal & Acctg	222	293	283	374	355	343	368	373	261	384	401	336	3,993	3,689
Marktg & Sales	456	602	580	767	729	703	754	764	535	787	823	688	8,188	16,238
Office Supplies	972	1283	1236	1636	1554	1498	1607	1629	1141	1679	1754	1468	17,457	15,639
Postage & Delivry	211	279	268	355	337	325	349	354	248	364	381	319	3,790	3,456

Figure 7.8: Example 2: Cash Receipts and Disbursement Journals *(continued)*

Facility Business Company

Cash Receipts and Disbursements Journals' Summaries

2/15/96

1995	Jan	Feb	Mar	Apr	May	Jun	Jul	Aug	Sep	Oct	Nov	Dec	1995 Y.T.D. 12 Mos	1994 12 Mos
Travel	254	336	323	428	407	392	421	426	299	439	459	384	4,568	7,995
Bus Entertainment	474	626	603	798	758	731	785	795	557	820	856	716	8,519	6,590
Auto Expense	292	385	371	491	466	450	482	489	342	504	526	440	5,238	4,569
Dues & Subscriptions	159	210	202	267	254	245	263	266	186	274	287	240	2,853	2,544
Equipment Rental	983	1299	1251	1655	1572	1516	1627	1649	1155	1699	1775	1485	17,666	15,345
Operating Supplies	1245	1645	1583	2096	1991	1919	2060	2088	1462	2151	2248	1881	22,369	20,996
Tele & Utilities	1200	1585	1526	2020	1918	1850	1985	2012	1409	2073	2166	1812	21,556	18,697
Rent	3025	3025	3025	3025	3025	3025	3158	3158	3158	3158	3158	3158	37,098	35,669
Insur (Prop/Liab)	792	1180	1180	1180	1180	1180	1180	1180	1294	1294	1294	1294	14,228	13,555
Computer Service	278	368	354	469	445	429	461	467	327	481	502	420	5,001	4,621
Interest	3286	3192	3099	3105	2810	2716	2757	2528	2435	2341	2247	2153	32,669	35,291
Income Taxes			1954			1953			1954			1954	7,815	0
Miscellaneous	536	709	682	903	858	827	888	900	630	927	968	810	9,638	12,456
DEPRECIATION INFO													57,946	45,814
General Expense	31844	40079	40724	48963	46575	47019	48029	48356	37899	49543	51361	45931	$536,323	$502,487
Total Disbursed	$51,577	$66,145	$65,822	$82,187	$78,127	$77,444	$80,677	$81,448	$61,078	$83,641	$86,983	$75,740	$890,869	$826,777
OPERATING INCRE(DECRE)	2586	5401	3067	9004	8482	6062	8935	9383	2542	9954	10795	6080	$82,291	$72,820
Debt Principal	3189	3283	3376	3469	3466	3560	3656	3749	3844	3939	4033	4126	43,690	41,698
Equipment Buys		4520	786	521		15669	5000	5663		3730			35,889	42,678
Cash-Beginning	10174	9571	7169	6074	11088	16104	2937	3216	3187	1885	4170	10932	10,174	21,730
CASH-ENDING	$9,571	$7,169	$6,074	$11,088	$16,104	$2,937	$3,216	$3,187	$1,885	$4,170	$10,932	$12,886	$12,886	$10,174

*Dates for all examples are fictional.

Figure 7.8: Example 3: Trial Balance

Facility Business Company			Trial Balance				11/15/95 Adjusted Trial Balance
1995	1994 CR & CD	1995 CR & CD	Accrual Debit	Adjustmt Credit	Adjustg Debit	Entries Credit	
Postproduction							
Editing	($420,189)	($472,077)	$35,500	($38,300)			($474,877)
Graphics	(162,231)	(173,132)	$19,900	($16,200)			($169,432)
Audio	(89,429)	(98,714)	$5,900	($4,900)			($97,714)
Other	(27,895)	(28,149)					($28,149)
Postproduction	($699,744)	($772,072)	$61,300	($59,400)	$0	$0	($770,172)
Duplication	(21,215)	(24,799)	$4,600	($2,000)			($22,199)
Production	(170,459)	(171,483)	$18,400	($19,700)			($172,783)
Interest	(2,170)	(1,767)					($1,767)
Other	(6,009)	(3,039)					($3,039)
Total Receipts	($899,597)	($973,160)	$84,300	($81,100)	$0	$0	($969,960)
DISBURSEMENTS							
Cost of Sales:							
Talent/Script	$18,945	$20,112					$20,112
Videotape	75,369	81,532	$6,300	($4,900)			$82,932
Production Expense	58,321	63,090	$4,900		$575		$68,565
Freight	21,865	22,634					$22,634
Travel	32,659	35,330					$35,330
Per Diem	16,899	19,621					$19,621
Art/Set Costs	6,490	6,969					$6,969
Location Fees	3,599	3,945				($575)	$3,370
Outside Services	85,643	96,445	$7,400	($11,600)			$92,245
Film Costs	4,500	4,868					$4,868
Total COS	$324,290	$354,546	$18,600	($16,500)	$575	($575)	$356,646
General Expenses							
Salaries & Wages	$241,847	$265,412					$265,412
Payroll Taxes	31,198	34,668					$34,668
Employee Benefit	12,092	13,597	$800				$14,397
Legal & Acctg	3,689	3,993					$3,993
Marktg & Sales	16,238	8,188					$8,188
Office Supplies	15,639	17,457	$2,100	($3,900)			$15,657
Postage & Delivery	3,456	3,790					$3,790
Travel	7,995	4,568					$4,568
Bus Entertainment	6,590	8,519					$8,519
Auto Expense	4,569	5,238					$5,238
Dues & Subscriptions	2,544	2,853					$2,853
Equipment Rental	15,345	17,666		($6,000)			$11,666
Operating Supplies	20,996	22,369	$2,300				$24,669
Tele & Utilities	18,697	21,556					$21,556
Rent	35,669	37,098					$37,098
Insur (Prop/Liab)	13,555	14,228	$2,800	($2,700)			$14,328
Computer Service	4,621	5,001	$1,800				$6,801
Interest	35,291	32,669					$32,669
Miscellaneous	12,456	9,638					$9,638
Depreciation	45,814				$57,946		$57,946
General Expense	$548,301	$528,508	$9,800	($12,600)	$57,946	$0	$583,654
Total Expenses	$872,591	$883,054	$28,400	($29,100)	$58,521	($575)	$940,300
Pretax Profits	($27,006)	($90,106)	$55,900	($52,000)	($58,521)	$575	($29,660)
Income Taxes	$0	7,815				($1,022)	$6,793
Net Profit	(27,006)	(82,291)					(22,867)
Retained Earnings							
Beginning							$919
Ending							($21,948)

Figure 7.8: Example 4: Journal Entries

		Adjusting Journal Entries	
Facility Business Company			11/15/95
1995		**Debits**	**Credits**
#1.	EDITING	$35,500	
	GRAPHICS	19,900	
	AUDIO	5,900	
	DUPLICATION	4,600	
	PRODUCTION	18,400	
	RETAINED EARNINGS		$84,300
	ADJUST 12/31/94 ACCOUNTS RECEIVABLE		
	RECEIVED 1995, EARNED 1994		
#2.	ACCOUNTS RECEIVABLE	81,100	
	EDITING		38,300
	GRAPHICS		16,200
	AUDIO		4,900
	DUPLICATION		2,000
	PRODUCTION		19,700
	ADJUST 12/31/95 ACCOUNTS RECEIVABLE		
	EARNED 1995, RECEIVABLE IN 1996		
#3.	RETAINED EARNINGS	29,100	
	VIDEOTAPE		4,900
	OUTSIDE SERVICES		11,600
	OFFICE SUPPLIES		3,900
	EQUIPMENT RENTALS		6,000
	INSURANCE		2,700
	ADJUST 12/31/94 ACCOUNTS PAYABLE		
	1994 EXPENSE, 1995 PAYMENT		
#4.	PRODUCTION EXPENSES	4,900	
	VIDEOTAPE	6,300	
	OUTSIDE SERVICES	7,400	
	EMPLOYEE BENEFITS	800	
	OFFICE SUPPLIES	2,100	
	OPERATING SUPPLIES	2,300	
	INSURANCE	2,800	
	INTEREST	1,800	
	ACCOUNTS PAYABLE		28,400
	ADJUST ACCOUNT PAYABLE 12/31/95		
	1995 EXPENSE, TO BE PAID 1996		
#5.	DEPRECIATION	57,946	
	ALLOWANCE FOR DEPRECIATION		57,946
	1995 DEPRECIATION EXPENSE		
#6.	PRODUCTION EXPENSES	575	
	LOCATION FEES		575
	ADJUST PRODUCTION EXPENSE		
	MISCODED TO LOCATION FEES		
#7.	PREPAID EXPENSES	1,022	
	INCOME TAX		1,022
	ADJUST FOR 1995 TAX PROVISION		

Figure 7.8: Example 5: Financial Statements

March 6, 1996

Facility Business Company Income Statements
As of December 31, 1995 and 1994

Statements of Income and Retained Earnings

	December 31, 1995		December 31, 1994	
SALES:				
Postproduction				
Editing	$474,877	49%	$435,189	48%
Graphics	$169,432	17%	$160,231	18%
Audio	$97,714	10%	$89,429	10%
Other	$28,149	3%	$27,895	3%
Postproduction	$770,172	79%	$712,744	79%
Duplication	$22,199	2%	$25,115	3%
Production	$172,783	18%	$160,669	18%
Interest	$1,767	0%	$2,170	0%
Other	$3,039	0%	$6,009	1%
Total Sales	$969,960	100%	$906,707	100%
COST OF SALES				
Talent/Script	$20,112	2%	$18,945	2%
Videotape	$82,932	9%	$77,369	9%
Production Expense	$68,565	7%	$62,321	7%
Freight	$22,634	2%	$20,864	2%
Travel	$35,330	4%	$37,659	4%
Per Diem	$19,621	2%	$17,899	2%
Art/Set Costs	$6,969	1%	$6,490	1%
Location Fees	$3,370	0%	$5,599	1%
Outside Services	$92,245	10%	$89,690	10%
Film Costs	$4,868	1%	$5,329	1%
Total Cost of Sales	$356,646	37%	$342,165	38%
GROSS PROFIT	$613,314	63%	$564,542	62%
OPERATING EXPENSES				
Salaries & Wages	$265,412	27%	$241,847	27%
Payroll Taxes	$34,668	4%	$31,198	3%
Employee Benefits	$14,397	1%	$12,092	1%
Legal & Accounting	$3,993	0%	$4,800	1%
Marketing & Sales	$8,188	1%	$17,238	2%
Office Supplies	$15,657	2%	$18,639	2%
Postage and Delivery	$3,790	0%	$3,456	0%
Travel	$4,568	0%	$9,224	1%
Business Entertainment	$8,519	1%	$6,590	1%
Auto Expense	$5,238	1%	$4,589	1%
Dues & Subscription	$2,853	0%	$2,544	0%
Equipment Rental	$11,666	1%	$16,345	2%
Operating Supplies	$24,669	3%	$25,998	3%
Telephone & Utilities	$21,556	2%	$19,679	2%
Rent	$37,098	4%	$35,669	4%
Property and Liability Insurance	$14,328	1%	$13,996	2%
Computer Services	$6,801	1%	$7,889	1%
Interest	$32,669	3%	$35,291	4%
Miscellaneous	$9,638	1%	$13,896	2%
Depreciation	$57,946	6%	$45,814	5%
Total Operating Expense	$583,654	60%	$566,794	63%
OPERATING RESULTS	$29,660	3%	($2,252)	-0%
Provision for Income Taxes	$6,793	1%	$0	0%
NET INCOME	$22,867	2%	($2,252)	-0%
Retained Earnings - Beginning	($919)		$1,333	
RETAINED EARNINGS - ENDING	$21,948		($919)	

Figure 7.8: Example 5: Financial Statements (continued)
Facility Business Company Balance Sheets
As of December 31, 1995 and 1994

ASSETS

	December 31, 1995	December 31, 1994
CURRENT ASSETS		
Cash	$12,886	$10,174
Trade Accounts Receivable	81,100	84,300
Prepaid Expenses	1,022	0
Total Current Assets	$95,008	$94,474
PROPERTY AND EQUIPMENT		
Furniture and Fixtures	$ 28,701	$ 13,435
Equipment	337,862	317,239
Leasehold Improvements	89,706	89,706
Property and Equipment - Cost	$456,269	$420,380
Allowance for Depreciation	($128,807)	($70,861)
Property and Equipment - Net	$327,462	$349,519
OTHER ASSETS		
Lease Deposit	1,000	1,000
TOTAL ASSETS	$423,470	$444,993

Liabilities and Equity

	December 31, 1995	December 31, 1994
CURRENT LIABILITIES		
Accounts Payable and Accrued	$28,400	$29,100
Current Portion Of		
Capitalized, Lease	45,782	43,690
Total Current Liabilities	$74,182	$72,790
LONG-TERM DEBT		
Capitalized Lease	$172,340	$218,122
Note Payable - Shareholder	30,000	30,000
Total Long-Term Debt	$202,340	$248,122
Total Liabilities	$276,522	$320,912
EQUITY		
Common Stock (1000 Shares Authorized;		
100 Shares Issued at $10 Par)	$ 1,000	$ 1,000
Paid in Capital	124,000	124,000
Retained Earnings (Deficits)	21,948	(919)
Total Equity	$146,948	$124,081
TOTAL LIABILITIES AND EQUITY	$423,470	$444,993

Figure 7.8: Example 5: Financial Statements (continued)

Facility Business Company		March 6, 1996
Statements of Cash Flows 1995 and 1984		
	1995	**1994**

	1995	1994
CASH PROVIDED BY:		
Operating Sales	$969,960	$906,707
Decrease (Increase) in Receivables	$3,200	($7,110)
Decrease (Increase) in Other Current Assets	($1,022)	$0
Cash Provided by Operations	$972,138	$899,597
Proceeds of Term Debt/Notes	$0	$0
Total Cash Provided	$972,138	$899,597
CASH USED FOR:		
Operating Expenses	$947,093	$891,591
Less: Depreciation - a Non-Cash Expense	($57,946)	($45,814)
Decrease (Increase) in Accounts Payables and Accrued Liabilities	$700	($19,000)
Decrease (Increase) in Current Portion Capitalized Leases	($2,092)	$0
Cash Used by Operations	$887,755	$826,777
Property and Equipment Expenditures	$35,889	$42,678
Reductions in Long-Term Debt	$45,782	$41,698
Total Cash Used For:	$969,426	$911,153
INCREASE (DECREASE) IN CASH	$2,712	($11,556)
Beginning Cash	$10,174	$21,730
ENDING CASH	$12,886	$10,174

Facility Business Company		March 6, 1996
Selected Key Financial Ratios 1995 and 1994		
	1995	**1994**

	1995	1994
- CURRENT RATIO		
Current Assets		
Current Liabilities	1.28	1.30
- DEBT TO EQUITY		
Total Liabilities		
Total Equity	1.88	2.59
With Shareholder Debt		
Treated as Equity	1.39	1.89
- RETURN ON TOTAL ASSETS		
Net Profit		
Average Total Assets	5.3%	N/A
- RETURN ON EQUITY		
Net Profit		
Average Equity	16.9%	N/A
- PROFIT MARGIN		
Net Profit		
Total Sales	2%	0%

offered, take them, even if you have to borrow. They are generally advantageous for the buyer. So if you are able, generally take a cash discount.

Timing of Vendor Payments   You can control when a specific vendor gets paid. As a rule, the faster you pay, the better the vendor will treat you. The reverse isn't always true, but is often the case. You can calculate how much interest you can earn, or not pay, by varying payment terms from 30 to 45 to 60 days. The resultant interest needs to be weighed against the cost-impaired vendor relationships and the internal hassle of increased C.O.D. demands.

As a short-term method of financing, you can stretch out your payments to your vendors to deal with temporary cash set backs. This is not an approved method, but is sometimes necessary. If the vendors have been with you through good times, they'll generally, if temporarily, work with you in slow times. Using your suppliers as bankers is a well not to be drawn from too often.

Priority of Vendor Payments   Suppose you are in a situation with your back to the wall—too many bills and too little cash and it is too early to pack it in. This has happened to the best of companies at least once in the up and down cycles of business. It is not necessarily pleasant, but it can be survived. You need to let your key vendors know what's going on, but first you have to have a survival plan. Internally, prioritize your payables files into must pay ASAP, then pay some when you can, to pay as little as possible until healthy. How you prioritize each vendor will be a function of how critical each is to your survival plan and the degree of their tolerance for not bringing legal action. Try to establish a getting current payment plan with each vendor—and do your best to live up to your promises. To minimize the negotiations, try to pay off all your small vendors, so you are dealing with a manageable number of vendors who really care about your survival. After all is said and done, your priority factors probably will be somewhat as follows: rent and utilities, critical suppliers, small vendors then everyone else. It is a difficult period to get through, but with some hard work, straight talk, and luck, it can be survived.

CASH FLOW MANAGEMENT

Cash flow management for the small- to medium-size business entails efforts to get money in faster and slow it down going out. Its reward in good times is the time/value of money—interest. Its reward in bad times is enough cash to survive until the good times return.

8 Financial Analysis

by James Spalding, Jr.

HOW IS YOUR BUSINESS DOING?

Do you know how your business is doing? "Compared to what" is a common refrain. By calculating your business' financial ratios for differing periods, you can compare your company to itself. If comparable industry averages are available, you can compare your ratios to the norm for your particular business segment. And if you just want to compare to businesses in general, the following definitions and discussion can give you some general comparison points. Remember, the more general the comparison, the less specific the possible conclusion. By comparing your business to itself at different times and to industry norms, while being aware of general business norms, you can best make use of simple financial analysis, ratios and percentages to monitor your business' health.

VARIANCE ANALYSIS—RATES AND VOLUMES

The first level of financial analysis is to compare the actual results with a predetermined standard of measure. Compare how the actuals vary from the standard. This is called **variance analysis.** The standard used could be last year's actual results, this year's budget or this year's average sales or expenses. How do the current actual results vary (the variance) from the standard used? Are the expenses up or down, by how much in absolute dollars and by what percent? What are the causes for the change from standard?

First, determine the absolute dollar variance, then calculate the percentage using the standard as the base. And then explain in words what happened and why. For example, we are using last year's actuals as the standard. Videotape expenses were $36,000 in 1989 and $48,000 in 1990. The variance is a $12,000 increase, a 33% change ($12,000/$36,000). This one is easy when we review our variance analysis on sales—videotape sales increased 36%. It makes sense that the related expenses would increase proportionally. Of course, the assumptions were shaped to make a facile solution. A little digging and thought can produce similar analysis with actual data. Do not expect an exact fit, just reasonable correlations.

The next level of variance analysis is to consider **quantity** versus **price** or **cost** variances. Quantity and price are implicit assumptions of a standard. For example, last year's videotape expenses represented 7,200 tapes sold at a cost of $5 each (7,200 × $5 = $36,000). Establishing quantities and pricing may take some research and averaging, but it can be done. We are looking for reasonable explanations, not detailed exactitudes. This year's expenses rose to $48,000. Our inventory system tells us that we sold 9,000 tapes, for an increase of 1,800 tapes. Spot checks in suppliers invoices show an average price of $5.35 for the year. So the quantity variance is actual quantity less standard quantity × standard price = (9,000 − 7,200 = 1,800) quantity variance @ standard price of $5 = $9,000 quantity variance. The price variance is a similar method—actual price versus standard price at standard quantity—$5.35 −

Table 8.1:   Quantity Price Variance

Quantity Variance	$ 9,000
Price Variance	2,520
Unexplained	480
Total Variance	$12,000

$5 = \$.35 \times 7,200 = \$2,520$ price variance. (See Table 8.1.) Remember to use the standard as the base number for cost and quantity. Note that exactitude is neither sought nor achieved. However, most of the increase is explained. And it is hoped, the quantity sold will be fairly close to the quantity invoiced the clients. But that's another analysis.

RATIOS AND PERCENTAGES FOR COMPARISONS

Financial ratios are somewhat abstract. They may be initially difficult to understand. Ratios are the numeric expression of a relationship between two differing numbers or data bases. They give you a perspective on the relationship, a sense of proportion. A ratio can be expressed as a pair of numbers (2:1) or as a percent (200%). For example, the relationship of 2 to 1 as a ratio is 2:1, or 200%. Financial ratios take this simple concept and apply it to your balance sheets and profit and loss statements. They are primarily used with accrual kept books, rather than cash basis statements. Financial ratios themselves are fairly straight forward; it's understanding the accrual and financial process that may take some work. Again, it generally takes some working with financial ratios to understand them, and then, more important to have financial ratio analysis improve your business' health. Four types of ratios will be discussed.

1. **Liquidity ratios** measure how able a business is at paying its bills.
2. **Assets management ratios** tell how well you are controlling and using your business equipment and property.
3. **Debt management ratios** monitor use of credit and debt.
4. **Profit ratios** measure bottom line impact and whether you'd be better off putting your money in a savings account. (To see examples of calculated Ratios see Example 5, Chapter 7, Facilities Business Company.)

Balance Sheets

Liquidity Ratios   A liquidity ratio tells you how able you are to meet your current debt obligations. It's a measure of solvency. In its simplest form, it's the old question of "do you have enough money to pay your bills?" For service companies, the acid test is to compare your cash to your soon-to-be-paid bills. Once this survival test is met on an ongoing basis, the current ratio is used to measure relative liquidity or your ability to pay your bills on time.

Current Ratio = Current Assets / Current Liabilities   Assets are things that you own. Liabilities are obligations you owe to creditors. **Current** in financial statement jargon means due as a receivable or payable within the next twelve months. **Current assets** include cash, receivables, inventories and prepaid expenses. **Current liabilities** include trade payables, accrued liabilities and that portion of long-term debt due within the next year.

The higher the current ratio, the larger the amount of current assets that is available to pay the due bills and obligations. A 2 to 1 current ratio is considered healthy; a 1.5 to 1 ratio is considered marginal; a 1 to 1 ratio, which many growing film and video businesses have, is considered risky.

Asset Management Ratios   How many days are your receivables outstanding before they are paid? How many times a year does your inventory turn over? How many sales dollars do your equipment and other fixed assets bring in each year? Asset management ratios answer these questions. They are the measure of management's efficiency in controlling and utilizing the firm's assets. Monitoring the ratios can help you improve the answers.

Days Receivables Outstanding = Average Receivables / Sales per Day
Sales per Day = Annual Sales / 360 days   This ratio gives you an indicator of how fast you are collecting on your sales. Forty-five to 60 days from the date of sale to day of cash receipt is not unusual when dealing with advertising agencies and large corporations and their related suppliers. Thirty to 20 days is excellent collection work. Sixty to 75 days means you should review to whom credit is extended, how fast and accurate your billing system is, and your post billing collection efforts.

Inventory Turnover = Annual Cost of Sales/ Average Inventory   This tells you how fast your inventory turns. The faster it turns, the better the return in sales dollars for each dollar invested in inventory. The lower the ratio, the more likely your inventory is not acceptable to your clients. Ratios of 3:1 to 5:1 show slow turnover, that is, you are only renewing or turning over your inventory three to five times per year. Look at your inventory (primarily tape stock and salable supplies) and determine what needs to be deleted; research what new hot selling items need to be added. High ratios of 8:1 to 11:1 indicate a pat on the back is called for since your inventory is being turned fairly fast.

Fixed Asset Utilization = Annual Sales / Total Fixed Assets This ratio measures the utilization of plant and equipment. A ratio of $4 to $5 of sales to each dollar of plant and equipment indicates a healthy usage of your fixed assets. $5 of sales per $1 of equipment is better than $1 of sales for the same equipment. A ratio of 1:1 indicates a capital intensive company that should review its level of investment in plant and equipment.

Debt Management Ratios These ratios monitor how able a company is in handling its debt. What is the relative mix of outside credit use versus the owner's equity investment? This is answered by the debt to equity ratio, which measures a firm's degree of financial leverage. How able is a company at handling debt? The debt service ratio measures the firm's ability to repay its existing and any proposed new debt.

Debt to Equity Ratio = Total Liabilities / Total Equity Here total liabilities include current payables and long-term debt. Equities include paid-in capital and retained earnings. This ratio tells you what the relative relationship is between the owner's investment and outside creditors. This measures financial leverage. A highly leveraged company is one with a ratio of two to three debt dollars to one equity dollar. For example, a company with $50,000 of trade payables, and $250,000 of term debt would have $300,000 of total liabilities. If their equity consisted of $75,000 of paid-in capital and $25,000 of retained earnings ($100,000 equity), their debt to equity ratio would be 3:1, a highly leveraged and risky company. Leverage is using other people's money to earn more for their equity money. In order to pull this off, the highly leveraged company has to earn more with the debt monies than they pay in interest. Leverage is a two-sided affair. When times are good, leverage improves your return on eq-

uity; when things go bad, leverage accelerates bad news. And then you still have to pay the debt money back. A debt to equity ratio of 1:1 and below indicates a conservative approach to using other people's money. Here equity financing is used either because of the temperament of the owners or as a result of the inability of the owners to obtain financing. A low debt to equity ratio is a safer approach, but many not maximize equity investment.

Debt Service Coverage Ratio = Net Income + Depreciation / Annual Principal Repayments This ratio is commonly used by lenders to see how able a company is at handling its existing debt requirements. A ratio of one dollar of existing cash flow (net income + depreciation) to one dollar of debt repayment shows a company with a tight cash flow situation. A lender might not want to risk its funds in such a business. A ratio of 2:1 indicates adequate cash flow to repay existing debt and maybe more.

INCOME STATEMENTS

Percent of Sales Analysis

The reason for all these ratios and monitoring is the proverbial bottom line. You need to determine what impacts the bottom line and how much profit there is from your investment. Profit ratios are the key measures of overall management ability.

The percent of sales analysis needs to be monitored within a company over time so that percent trends can be established. (See Table 8.2.) When you compare your company to another or to industry averages, differing accounting treatment of similar items can cause significant percent differences. Be careful. If you are using another company's percentages, eliminate accounting differences and then analyze and conclude. For your own percentages, compare them against prior trends and analyze significant changes.

Table 8.2:   Percent (%) of Sales Analysis

Percent of Sales Analysis = Cost or Profit Line Item / Sales	
% Cost of Sales	= Cost of Sales / Sales
% Overhead Expense	= Overhead Expense /Sales
% Interest Expense	= Interest Expense /Sales
% Net Profit	= Net Profit / Sales

Return on Investment = Net Income / Equity   The net profit for a sole proprietor or partner should be after an allowance for fair compensation for services rendered. Wages for work performed come out before profits are determined for the equity monies invested. For corporations, the net profit number used is the after tax amount, that which is available to be retained in the business or paid out in dividends.

The return on investment (ROI) percent is the measure of how much the owners have earned on their equity. A 15% to 20% return is fair for a well-run small business. A 30% plus return is doing the right thing in the right place at the right time. A return of 10% and under suggests you might be better off putting you money in a savings account.

Break-Even Analysis

What sales volume is necessary to break-even? That is, what sales level produces no profits and no losses, so that any additional sales will produce profits? This is called the break-even analysis. To calculate it, you must know what your gross profit percent is and the amount of your fixed or overhead expenses. Assume your gross profit percent is 40%, based on last year's sales of $800,000 and a cost of sales of $480,000, resulting in a gross profit of $320,000 or 40% of sales. This year your overhead expenses are $400,000. Given last year's gross profit margin of 40%, what sales volume is needed before the business begins showing a profit?

Break-Even Sales = Fixed or Overhead Costs / Gross Profit %
Break-Even Fixed Costs = Sales × Gross Profit %   Those of you who remember your algebra know that these equations are the same, just solving for a different unknown. If you have determined what your overhead costs are, what sales level is needed to break-even? Here fixed costs are $400,000 with gross profit percent at 40%, requiring a $1,000,000 sales level before profits begin. Conversely, if sales are given at $1,000,000, with a gross profit percent of 40%, this would leave $400,000 available for overhead. (See Figure 8.1.) The break-even analysis is very handy for "what if" analysis and budgeting. If you project a certain overhead level, what do sales need to be to break even? If you know your sales level, how much overhead can it support. And, of course, everything is better if you can increase your gross profit percentage.

The above financial ratios are tools of analysis. A ratio that works for one company may not work for another company. However, if your company's ratios don't fit the norm, there should be a good explanation, or consider it a warning sign. Usually financial ratio analysis provides meaningful insights into how your business is doing. By monitoring the ratios over time, these insights can result in a better run and a more profitable business.

GOVERNMENT REQUIREMENTS

Payroll and Related

Independent Contractor or Employee   An employee works for an employer. An independent contractor is someone who contracts with others outside the employer-employee relationship to provide goods and services. Pretty straightforward stuff at either end of the gray scale, but in the middle, who is what is not always so clear. For the hiring employer, the choice has different tax, insurance, government reporting and liability exposure. For the individual the choice means different personal income tax implications, control of working conditions, fringe benefits and, sometimes, degree of job security. We will take a look at what distinguishes an independent contractor from an employee and what are the implications of each choice for the employer and the individual.

Definition, Principles, and Determining Characteristics   If you don't always understand the difference between independent contractors and employees, join the ranks of the IRS, the state's employment development department and, of course, Congress and the courts. This can be an exceedingly gray area of the law. It is often resolved in real life by a payroll tax audit. The audit can result in sizable collections for the government after they tell you what their very "clear definition" is for your business. The IRS tells us that an employer/employee relationship "exists when the person for whom the services are performed has a right to control and direct the individual who performs the services, not only as to the results to be accomplished by the work, but as to the details and means by which that result is accomplished" [Treasury Regulation 31.3401(c) −1]. There are five determinant principles, followed by determining characteristics of the independent contractor and/or employee. In clear-cut

Figure 8.1: Break-Even Analysis

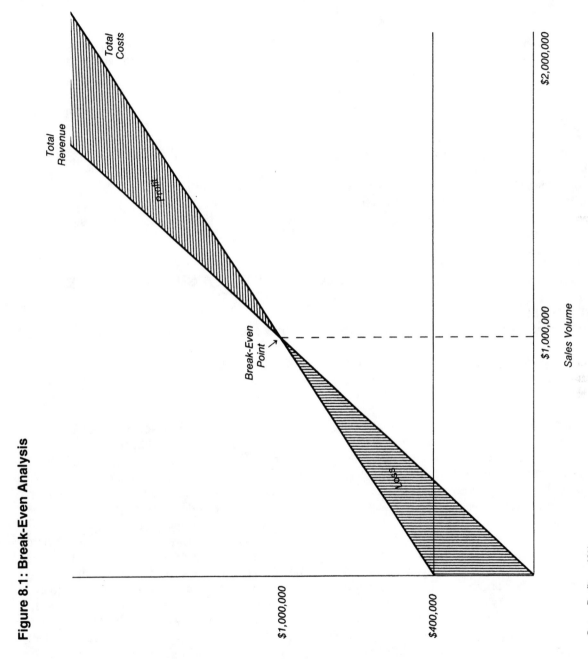

Total Costs

Total Revenue

Profit

Break-Even Point

Loss

$1,000,000

$400,000

$1,000,000

$2,000,000

Sales Volume

Gross Profit = 40%
Fixed Costs = $400,000

cases, it is easy to tell the differences between the two—so we think. But, often in the world of film and video, with frequent use of freelancers, the gray areas need some clarification.

Determining Principles

Right to Control   Does the contracting company control only the desired results—the contract specifications—and not the manner and means to achieve the sought after results? An independent contractor controls how the service or product is produced and delivered; an employee is ultimately controlled by the employer.

Independent Judgment   Does the individual exercise her own independent judgment in determining what tasks are to be performed, when and how, as well as creative and qualitative judgment? The independent professional is engaged for her expertise and not significantly second-guessed in execution of her expertise. While she is keeping the client happy, she is calling her own shots.

Profit Basis   Does the individual receive her pay in the form of profits—net income after expenses—rather than wages—hourly pay times hours worked. With the profit criterion, the freelancer determines her work methods and has independent expenses, including overhead and other job-related costs. These may include the engaging of other independent contractors and employees. And finally, the independent will normally provide the tools of her trade at her own expense. The independent contractor's recompense is net income or profits after billing the client for services and expenses and paying for all her costs of doing business. The employee's pay is his wages with out-of-pocket expenses being reimbursed by the employer. The independent contractor can incur a loss when receipts fall short of expenses. The employee may be underpaid, but cannot incur a business loss.

Contract Responsibility   Is the individual responsible for her own actions and the delivery of the end product or service? The contract part of this criterion is often oral, not formally written down, between the two parties. But the substance of the freelancer's contractual responsibility needs to be clear. The contractor is paid for the delivery of services or products; nondelivery would be a breach or the breaking of the contract and make the contractor liable for any possible damages. An employee is paid for his time, not

for his results or lack thereof, and, in fact, could make the employer liable for the employee's own actions while acting as an employee.

Degree of Supervision   Does the individual determine her own hours and methods and means of delivering her services or product? This is similar to the right to control, but with a difference that merits emphasis. An independent contractor is her own boss; an employee is supervised.

Whether an individual is an independent contractor or not is determined on a case-by-case basis. No single one of the above criteria is solely determinant. All factors need to be weighed with a good bit of common sense. It is important to analyze the characteristics of each individual to determine whether he or she is an independent contractor or employee. (See Table 8.3.) Not all characteristics will apply to all situations. Again, it is a judgment call, weighing the preponderance of facts to give the most reasonable results.

PAYROLL TAXES AND RELATED EXPENSES

It is generally more expensive for an employer to hire an employee than an independent contractor because payroll taxes must be added to the wages of an employee, but not to the independent contractor's bill. The independent contractor is responsible to the government for her own taxes, while the employer is responsible for employer taxes and the proper withholding of employee taxes. The payroll taxes and the related paperwork and responsibilities are the primary reasons for employers to prefer independent contractors to employees. Besides payroll taxes, fringe benefit packages such as vacation and sick pay, group medical and dental, pensions and the various other benefits increase the employer's costs. And finally there is the liability for the employee's actions to be concerned with in considering to hire an employee or independent contractor. Taxes, benefits and liability add up. But, this may be a small price to pay to own your employee's soul for 70 to 80 hours a week.

For the individual, the tables are turned. Freedom of operation—whether to work less or, often, more—is yours as an independent contractor. However, while the employer's taxes are high, as a self-employed individual, your taxes will be higher than as an employee. The main bugaboo

Table 8.3:   Characteristics of Independent Contractors and Employees

Independent Contractors	Employees
• Business letterhead and invoices	• Employer letterhead and invoices
• Business office and telephone	• Employer office and telephone
• Bills client for services rendered	• Time sheets submitted to supervisor
• Legally accountable for work performed	• Employer legally accountable
• Schedules own work hours	• Employer determines hours
• Own tools of trade	• Employer supplies tools
• Hires and terminates assistants	• Employer hires and fires
• Advertises services or product	• Employer markets company's services and product
• Multiple clients	• Usually one employer
• Federal Employer ID #	• Employee's Social Security #
• Own workers' comp and medical insurance	• Employer's workers' comp and group medical insurance

here is your friendly Social Security (FICA) tax. The employer matches the employee's 7 + % "contribution" to the old-age fund. Thus, as an employee, your effective contribution rate is 15 + %. As a self-employed person, you pay both shares, the employee's for yourself and the employer's for hiring you.

There are two other taxes to be mentioned— unemployment taxes paid by the employer and disability insurance paid by the employee. These vary from state to state. They are not as onerous as Social Security but need to be considered.

All the employer/employee related taxes, not income taxes, have a ceiling or maximum income or earnings level. For example, for 1991, both the payroll and the self-employment tax applies to the first $53,400 of income at 15.3%, then to $125,000 at 2.9% then ceases. Unemployment taxes are on the first $7,000 of income; disability insurance is on the first $25,149.

Table 8.4 displays the payroll and self-employment taxes for 1991. Given are the tax rate as a percent (%), the maximum earnings level for the tax rate (Max $), and the maximum tax (Max Tax)—the percent times the maximum earnings level.

The rates and ceilings can vary annually, and usually do and, of course, the state taxes vary from state to state. Check these annually.

A few separate words about the payrolling tasks. This is generally only a significant burden for the first employees. If you haven't had the privilege of meeting payrolls, making tax deposits and filing government reports, then you have another world awaiting you. After the first several employees, however, additional employees do not significantly add to the administrative tasks. A highly recommended way to minimize your record keeping duties is to engage an outside payroll service.

Employer's must also be aware of their liability with respect to an employee. Any action of a company employee while on, to and from company business is the responsibility of the employer. Independent contractors are legally responsible for their own actions and those of their employees.

Another difference between employees and independent contractors is unionization. Employees can form and join unions; independent contractors cannot, at least not your company's union. We have generally been using "freelancer" to describe an independent contractor; "freelancer" is not a legal term, but it can be applied to independent contractors and interim and part-time employees. The distinction here is between an independent contractor, who may be called a freelancer, and an employee, who may be called a freelancer, or full-time and part-time staff. The employee can join a company union, the independent contractor cannot. Independent contractors can form unions. The plumbers' union, the barbers' union, the writers' guild are examples. But you can legally choose not to engage their members unless you sign the union contract. Employees can form and join a union, and you are required to work with them and bargain in good faith with their representatives.

Two final points. 1099—MISC forms are required to be filed for all independent contractors paid $600 and over in any calendar year. A 1099 is similar to the W-2 the employer files on the employee, except it requires less information—just the freelancer's name, address, Social Security number and annual amount paid. And finally, the government has put teeth in its regulation to in-

Table 8.4: Payroll and Self-Employment Taxes, 1991

	Employer Tax:			Employee Tax:			Combined Tax:	
%	Max $	Max Tax	%	Max $	Max Tax	%	Max $	Max Tax
Federal: Social Security/FICA - OASDI + HI (Old Age & Survivors Disability Insurance + Hospitalization Insurance- Medicare)								
7.65%	$ 53,400	$4,085	7.65%	$53,400	$4,085	15.3%	$ 53,400	$8,170
- HI	$53,401	to		$53,401	to		$53,401	
1.45%	$125,000	$1,038	1.45%	$125,000	$1,038	2.9%	$125,000	$2,076
Unemployment Insurance								
.08%	$7,000	$ 56		None		.08	$7,000	$ 56
California: Unemployment Insurance								
3.4+ – %	$7,000	$238		None		3.4%	$7,000	$238
Disability Insurance								
	None		.9%	$25,149	$226	.9%	$25,149	$226

1991 Maximum Payroll Tax on Maximum Earnings of $125,000
Employer: $5,417   Employee: $5,349   Combined: $10,766
                          Maximum Tax - Employee - $10,766

Self-Employment Tax - FICA - OASDI + HI (Old Age & Survivors Disability Insurance + Hospitalization Insurance - Medicare)								
	None		15.3%	$53,400	$8,170	15.3%	$53,400	$8,170
- HI	None			$53,401	to		$53,401	
			2.9%	$125,000	$2,076	2.9%	$125,000	$2,076

Maximum Tax - Self-Employed - $10,246
Less Deduction Allowed @31% - Net Max Tax $7,070

sure that employers are being responsible in the classification of employees and independent contractors. If an individual is found to be an employee after having been treated as an independent contractor, the employer is liable not only for the employer's taxes, but also for the employee's withholding taxes, including Social Security and income taxes. Not much of a problem for a company that is utilizing true independent contractors who are honest tax paying citizens. An offset is allowed if the so-called "employee" filed and paid his taxes as an independent contractor. But if the company is wrongfully evading payroll taxes and hiring employees who are not having their taxes withheld, and then not reporting their income and thus also evading their income and Social Security taxes, then the company has a problem. How are you to know? Follow the previously noted principles and characteristics to differentiate in each case. If you are still in doubt, consult your tax or legal advisor. If you are doing the hiring, it is your job to know, or be willing to pay the consequences during a payroll tax audit.

INCOME TAXES

Besides keeping information for your own purposes, the government has also set requirements for you to follow to be able to take tax deductions. Deductions are important because they decrease your taxable income and thereby minimize the tax you pay. The more deductions, the lower the tax. The general goal of most taxpayers is to pay their fair share but no more than is necessary. Knowing and following the government's rules helps you pay less in taxes.

Business Versus Personal and Year-Paid Deductible

The government is concerned that the taxpayer does not mix nondeductible personal expenses with deductible business expenses. A business expense is an expense incurred for a profit motive in an active trade or business. The tax code has two general requirements for business deductibility.

The first is that the expense is **ordinary** and **necessary. Ordinary** is defined as a common and accepted item, not a capital expenditure; and it is a recurring or usual item. **Necessary** is defined as helpful or appropriate. To be deductible as a business expense, it must be ordinary and necessary to the particular business. This leaves a lot of leeway in its interpretation, and rightly so since what makes sense for one type of business may not make sense for another. If the expenditure could be judged as personal, ask yourself if it is ordinary and necessary to the business. Substantiation requirements for commonly abused deductions such as travel and entertainment will be discussed later, but the general rule of ordinary and necessary is fairly straightforward and applies readily to our everyday business decisions. The second requirement for deductibility is that the expense be **paid** in an active trade or business. For most businesses this means that you can't take a deduction until the bill is paid. Partial payments are also deductible in the year paid. The exceptions to the year paid rule are capital expenditures and prepayment of interest and insurance. Capital assets are written off as depreciation over their useful lives, which is five to seven years for most equipment purchased. Interest and insurance are deductible on the accrual basis, that is, when earned by the lender and insurance company. Expenses paid for with credit cards are deductible in the year charged. If you are an accrual basis taxpayer, the expenses need to be properly accrued.

Adequate Records

The government is concerned that you are able to substantiate your numbers. Estimates and maybes are not acceptable. It is important to keep accurate business records. In areas of frequent abuse, additional documentation is necessary. For cash receipts, you are expected to maintain a list of all payments from clients. This cash receipts journal is supported by client billings and deposit slips. If you receive actual cash, that's income also and should be recorded just as you would record other receipts. For most expenses, a listing by supplier with an amount that is supported by an invoice is good enough. For receipts and expenses, a separate business checkbook used exclusively for the business is a good idea. That's a sketch of basic business records. Records should be kept for a minimum of three years from the date your returns are due including extensions.

You would be safer to keep them five to seven years.

Substantiation

Besides the basics, the government requires specific substantiation of travel, entertainment including business meals, cars and trucks, personal computers and cellular communication devices used for business at home, and gifts.

Travel Travel is defined as an away from home over night trip. The expenses include transportation (to and from, and at the location), lodging, meals and incidentals. To substantiate these travel deductions, you need to record the amount of each expenditure, the dates and location, the business purpose and the business relationship. Try to keep all the receipts (required for the lodging receipts and any single expense over $25). Then write the client's name and the business purpose on each receipt. You may then summarize each trip by type of expense— transportation, meals and so on. Business meals while traveling are only 80% deductible. Keep a separate tally of business meals and multiply the total by 80% to arrive at the deductible amount.

Entertainment and Business Meals This is an area of abuse that the tax auditor reviews. That doesn't mean don't take a deduction; it just means do your homework. Here you need to note the amount, time, place and type of expense (meals, drinks, and so on), business purpose and business relationship. Entertainment and business meals must have detailed substantiation and, like travel meals are only 80% deductible. For both travel and business entertaining, you are usually dealing with present or prospective clients. Get a receipt and write on the receipt the client's name and business purpose. Bring these back to your office and summarize weekly (at least monthly).

Cars and Trucks For autos and trucks used for both business and personal use, the portion of business usage needs to be established. While a diary is not required, you somehow need to justify the business portion of the usage. A diary kept for several typical months out of the year is very useful. What you need to arrive at for a tax deduction, is the percent of **business versus personal** usage to allocate vehicle expense—depreciation, repairs, gas and insurance. Any vehicle must meet the 50% business usage test to qualify for the five-year write off under the accelerated method permitted by the tax code. If usage fails the 50%

test, the vehicle must be depreciated on a straight line basis (a certain fixed percentage each year) over five years. When recording your mileage, remember that commuting to and from work is personal, non-business mileage. If you use your car in business very little, you may record only your business mileage and take a deduction at the rate of 28.0 cents per mile (in 1992). The mileage rate is updated annually.

Luxury Cars Luxury autos are defined as cars that cost more than $12,800. The luxury car rules have annual dollar limits for depreciation (ACRS/ MACRS).

Employee Reimbursement Plans If an employer reimburses the employee for business auto usage, under an **accountable plan,** then the employer does not have to include the mileage reimbursement in the employee's W-2 wage reported to the IRS at the end of the calendar year. This simplicity of non-reporting is ample incentive for such an accountable plan to be used. An accountable plan requires the following:

- written business connection/purpose of the mileage,
- substantiation of the use, business purpose, the amount and time/date
- the return of excess advances in a reasonable period.

The IRS states that the reasonable return period include the following:

- the fixed date method—an advance no earlier than 30 days of the expense, substantiation within 60 days, and the return of excess advances within 120 days of the expense
- the periodic statement method—at least quarterly accountings and return of excess advances within 120 days of such accountings.

A **non-accountable plan** is any plan or non-plan that does not follow the accountable plan rules above.

Therefore, if you use the standard mileage rate method with your employees, it is all fairly straight forward. But if you have a non- accountable plan, you report the full amount of the reimbursement on the W-2 and withhold on the amount paid. And any excess advances, or funds not accounted for in a reasonable period are similarly reported via the W-2. And if you reimburse

at a rate that exceeds the IRS rate of 28.0 cents, the excess reimbursements are also included as reportable wages. For the employer, the amount paid in excess of the allowable mileage is deductible employee compensation. For the employee, the income is treated as wages. The employee may be able to take a miscellaneous employee deduction subject to the 2% of adjusted gross income disallowance.

Personal Computers For personal computers kept at home and used for business purposes, a diary of business and personal usage is required for tax deductions. As with cars, there is a 50% test for eligibility for the accelerated depreciation under the permitted methods. Business usage not in excess of 50% requires five-year straight line depreciation for the allocation between deductible business usage and nondeductible personal usage. If the computer is kept at your business location, then you can use accelerated depreciation without a diary.

Gifts Business gifts require the substantiation of amount, date and description, business purpose and the name and business relationship of the person receiving the gift. There is a $25 limit per gift on the amount deductible.

Estimated Taxes

Self-employed individuals must make **estimated tax** payments. Essentially, you estimate what your income will be for the year and also your deductions. Then calculate your estimated tax. Forms 1040ES for the Feds and similar state forms are used. If your estimated liability exceeds $500, then you must make quarterly installments on April 15, June 15, September 15 and January 15, or the Monday following if these dates fall on a weekend.

The estimated payment for the IRS includes the Social Security Self-Employment tax (see Table 8.4.). No tax applies if your self-employment income is under $400. If you are employed, the gross income from employment is the base. Then you add your self employment income on top. This is so you don't pay the same tax twice, once as an employee and once as a self-employed person. As an employee, the rate is not as high since your contribution is matched by your employer.

Filing

To file, you first need to summarize your accounting records, and of course your other tax

Table 8.5: List of Tax Forms

1040 U.S. Individual Income Tax Return
1040 Schedule A—Itemized Deductions
1040 Schedule B—Interest and Dividend Income
1040 Schedule C—Profit or (Loss) from Business or Profession
1040 Schedule SE—Computation of Social Security Self- Employment Tax
4562 —Depreciation and Amortization (including Section 179 election to write off up to $10,000 of capital business expenditures made during the year, if total capital expenditures were $200,000 and under)

records. Then the various forms need to be completed. (See Table 8.5.)

Returns are due April 15th, but extensions for filing are available through August 15th and, if needed, to October 15th. The actual taxes owed for the year are due April 15th regardless of extensions. Late-payment penalties apply to the underpayment of the total tax due if not paid in full by April 15th.

Underpayment Penalties

Penalties for underpayment of withholdings and estimated payments of the final tax liability are really interest charges. Use Form 2210 to make the calculation. It is a finance charge on the underpayment of the tax. Two easy exceptions to the penalties are paying at least last year's amount of tax, or 90% of the current year's tax. By April 15th, the total tax is to be paid. If underpaid, late payment penalties apply to the unpaid portion in addition to the interest penalties. If an extension is not filed, or the return is not filed on time, late filing penalties apply.

Audits

Audits are the IRS's threat to encourage taxpayers to comply with the rules of our "voluntary"

system. The IRS actually audits a minuscule percentage of returns. When they notify you, it is often to clarify one or two items on the return. These are sometimes cleared up with phone calls and mailed in copies of documentation. The IRS also audits a small percentage of returns chosen randomly. If yours is chosen, you must be prepared to show all your records. However, if they think you are cheating, they can be tenacious. They can also assess large penalties. Again these are for tax abusers. In most cases of honest mistakes, they just want the underpaid tax and finance charges. If they find you have overpaid your taxes, they'll issue you a refund check with interest.

Despite all the recent simplifications, in fact because of them, the tax codes are a maze of hidden traps. Nevertheless, many folks can still do their own returns. Helpful freebie guides for the do-it-yourselfers are IRS publications #334 "Tax Guide for Small Businesses" and #17 "Your Federal Income Tax." Both are available free of charge from your local IRS office. They make excellent reading just before bedtime. If your business affairs are more complex than simple, it often pays to consult a tax advisor or have a professional prepare your returns.

Other Requirements

Most all other levels of government have some form of taxes to feed the beast. Sales taxes are levied on sales incurred. Property taxes are assessed on real estate and, often, business and/or personal and investment properties. Some states and municipalities tax payrolls. Call your state income, franchise and equalization bureaus and inquire regarding additional taxes. Call your city hall and county seat and also inquire as to their particular needs. It is your responsibility to file tax returns and pay your fair share.

9 Obtaining Money

by James Spalding, Jr.

Obtaining a loan is one of life's Catch-22s. When you desperately need money, no one will lend you a cent. When your bank accounts are spilling over, you are besieged with bank officer's lending proposals. It is a fact of the business world. What this tells us is to establish lending relationships during good times and hope they will stay with you through the downturns. This chapter reviews the various lending terms and sources available, then the contents of a loan package and finally when to lease and when to buy.

GENERAL RULES OF FINANCING

There are some widely held generalities about financing. Long-term assets should be financed with long-term debt; working capital needs should be financed with short-term debt. What this means is that the life of the underlying assets should exceed the term of the debt. Don't buy a car that will last three years with a note that runs ten years. You don't want to be paying on a car that is long gone. The lender won't want to have as collateral something that is worthless before the debt is paid in full. Similar thinking applies to short-term debt arrangements. Use receivables, inventories and other assets to finance short-term cash shortages. If you don't have enough cash to carry your receivables or inventories, make a deal to borrow against your receivables and inventories. Then pay the loan down when the inventories are sold and the receivables are collected. The term of the debt has to fit the reason for the loan and the life of the underlying collateral. Again, use long-term debt for long-term assets; short-term debt for short-term assets.

The next point to note is the conservative nature of lenders. The lender is investing either his own money or other people's money. He wants it paid back with interest and without hassles. The collateral, cosigners, guarantors, and so forth, are used to cover his bets. He doesn't want the headache of selling your house or whatever other collateral that has been provided. You financier does not want to come in and run your business. His business is lending money to credit worthy debtors and earning loan fees and interest above what he has to pay for the money that he is borrowing to lend to you.

It is important to shop your loan proposal. Bankers and other financing sources can be arbitrary. One likes blue and the other red with no apparent rationale. If you try enough different sources, one may just be waiting for your proposal. Or you may be able to do some significant comparison shopping and bargain for better terms.

The final general rule of finance is a derivative of the first rule—when you need money the most, it is the most difficult to get. Seed money for a start-up is the most difficult to raise and the most costly in terms of interest and equity. Second round financing—given a good track record—is easier and less costly. Well-run, mature companies can pick and choose when and how to seek financing. The less you need it, the easier it is to get it; the riskier the deal, the harder the sell.

TERMS

Financing terms have generally infiltrated our general vocabulary.

Collateral—This is an asset that is pledged to the lender as a secondary source of repayment should the primary payment source not repay the loan.

Cosigner—Someone who signs the note along with the primary debtor and will pay the debt in the event the primary debtor fails to do so.

Default—Not fulfilling the terms of the loan agreement. A loan default can be corrected or waived by the lender. But a default can also allow the lender to call the loan. To call the loan means that the lender can require immediate and complete payment of the loan balance, or sell the collateral.

Fixed rate loans—The percent interest charged during the life of the loan is fixed over the life of the loan (not changeable) at the time the loan is made. Fixed-rate loans are generally at higher rates than variable rate loans to compensate for the lender's risk that interest rates may go higher. If rates go substantially lower, the borrower can usually refinance.

Guarantor—Someone who ultimately will pay the debt if all else fails.

Points—One point is equal to one percent; two points are equal to two percent, and so on. Points can be used to refer to the interest rates above the prime interest rate—two points above prime. If prime is at 10% then two above is an interest rate of 12%. Points are also used to calculate the loan fees of the lender. A loan fee of two points is equal to 2% of the loan amount. Thus with a $100,000 loan, a two point fee is equal to $2,000.

Prepayment penalty—This is a fee charged by the lender if the loan is paid down before the term has run out. You may be able to get better rates elsewhere, but the amount of the prepayment penalties could prohibit a move. Check the prepayment penalties—if any—before you sign original loan documents or refinance documents.

Prime—This is the interest rate charged by the leading banks to their best customers. It fluctuates up and down with the economy and is used as the base interest rate for variable rate loans.

Term—The life of the loan per the loan agreement.

    Short Term—One year and under in time.

    Long Term—Over one year in time.

Unsecured—A loan that it is made only on the good faith and credit of the borrower, without collateral.

Variable rate loans—Loans with interest rates that vary over the term.

SOURCES OF FINANCING

Sweat Equity

While sweat equity has not been formally accepted as a term in the lexicon of finance, it goes a long way to describe the start-up funding of many companies. We are talking about the unpaid and underpaid time and the use of the equity you may have in your house and investments, your borrowing ability with friends and relatives, and the ever increasing credit limit of your Master-Card, Visa, and other credit cards.

When you first start out, most any source is acceptable funding. For friends and relatives, structure a business deal that allows them to win while you are using their money. A debt format is better for the lender if the business fails—the debt can be written off in full in the year the debt goes bad. An equity format limits the write off to an annual $3,000 limit for all net capital losses. (However "small business corporation stock" that qualifies under IRS Section 1244 allows an exception to this capital loss rule. $50,000 or $100,000 on a joint return, can be written off as an ordinary loss for this special 1244 qualifying stock.) You also can informally offer future equity participation should the business prosper. Since you are dealing with friends, and especially with relatives, put the debt in writing. This is important to make the deal a business deal for both sides and to provide documentation for the IRS to demonstrate that the lender—your pal or aunt—entered into the deal to make money. A gift or donation to you is not a business or charitable write off. It may be your best form of financing, but it gives the donor no write offs.

Another form of financing is to make deals with people you need to make the business work. Exchange their efforts for a piece of the business or arrange a deferred payment schedule. This can be done within a sole proprietorship format by promising a percent of future revenues or a piece of the bottom line. A percent of a partnership or corporation also would work. (Consult a tax advisor to avoid income in the year of contribution—IRS Section 83 Elections)

Limited Partnerships

A limited partnership is one that "limits" the limited partners' role and risk in the business to their money. A general partner runs the business and bears the burden of general liability. The limited partner is at risk only for his investment in the business and has no role in the management of the company other than selecting the general partner. A limited partnership can be an "informal" private offering among friends and associates, or a public offering through investment bankers and registered security brokers. Once you get beyond a small circle of friends you need to contact a lawyer who specializes in securities law and offerings. Because of the untold number of scams that have been perpetrated on the investing public, the federal Securities and Exchange Commission, and the state's version of the same, have stringent rules about private and public offerings. Specific legal advice for a small private placement would be time and money well spent. For a public offering, the legal formalities would be handled by the investment banker with broker connections—you may have to pay for the lawyers and assist in the formation of the offering document, but you can follow their expertise with your lawyer's review. The trick about a public offering is not the legalities, they will get done, but having your business plan accepted by an investment banker in the first place.

With the present Passive Activities Loss (PALS) limits, limited partnerships are not as attractive as they were under pre-1986 laws that allowed tax loss write offs to the individual investor. Now, a passive investor must carry her losses until there are offsetting gains in this investment or other passive investments, or she may take the losses when she sells the investment.

Limited partnerships can be simple—such as with a small number of investors who are familiar with you and the business. When kept simple, they can be a funding source for start-ups and projects that can be spun off from a small company while providing investment opportunity for the limiteds. When the funding needs escalate, so do the complexities and fees of doing the deal.

Venture Capitalists

These are the mythical people of the financial world who give you money, when you have a good business idea. Well, they are not really mythical, but per the first rule of finance, they only give money for good ideas to those that don't need their money. A venture capitalist who provides initial funding for a start-up is rare. This occurs when the start-up's CEO has already made the venture capitalist rich in a former business. Normally, a venture capitalist invests in stage two funding—after your sweat equity has proven that the business has a reasonable chance for success. To work, you need a business plan, a telephone, and patience. Venture capitalists can be found by going to your local business library and reviewing the various books under the venture capitalist category. Venture capitalists generally want a significant chunk of the business—20% to 50% and some form of liquidity within a reasonable time frame—five to seven years. The typical liquidity would be to bring the company public or to sell to another concern. An experienced venture capitalist who has invested in your company can be an excellent source of management expertise to help you through the hurdles of a start-up to young growth company.

Stock Offerings

This source of money is usually available only to companies that have a track record. However some start-ups are funded with public offerings. Private stock offerings are what you offer to friends and venture capitalists. They need to be done properly and kept private by conforming to the securities laws (see your securities lawyer). Public stock offerings however are a specialty in themselves. As with a public offering of a limited partnership, you first need to convince an investment banker. Once done, the banker's and their counselors can guide you through the maze of SEC and state laws. You probably will have to give up a significant chunk of your business, but you probably also will be getting very rich, and able to afford the resulting headaches and considerable expenses of a public company. Again public offerings are not the norm and are limited to companies with successful track records or the promise of breakthrough technologies.

Equipment Financing

This can often be the most easily obtained financing. Many times the vendor, eager to make a sale, will arrange the loan or lease for qualifying buyers. You just need to qualify. This normally means several years of profits, or, if your personal credit can bear the burden, cosign or guarantee the loan for the business. Normally, the owners of a small business have to personally guarantee the business' debts in any event. With equipment financing, the equipment is the collateral for the loan. The term is for a period less than the life of the equipment. In addition to vendor, and vendor arranged financing, banks, leasing companies, and brokerage firms will finance equipment.

Receivables Financing

Receivable financing is using your accounts receivable for collateral in a debt arrangement. A variation of this is called receivable factoring, that is where you actually sell specific accounts to a "factor" who does the collection work and keeps the receivables proceeds. Receivable financing can be a good source of money for a young company. Typically the lender advances monies against 70% to 80% of your receivables not over 90 days old. Just the good accounts are eligible for collateral. For a service business without inventory requirements, receivable financing gives you higher debt capacity. Thus, when sales levels are high, receivable financing provides additional monies to pay the additional bills that supported the higher sales. The interest rates can be fairly steep—up to 6% and 7% above prime. And usually at the higher rates, the risks to the lender are greater. The lender often requires a copy of each receivable to be physically delivered to them as specific collateral. Receivables financing is money at high interest rates and high administrative costs. But sometimes expensive money is better than none at all. The factoring part of this type of financing occurs when you actually sell your receivables to the factor and they do the collecting of their recently purchased receivable. In this type of arrangement, the seller will receive 60% to 90% of the stated value of the receivable. Quite a discount! On the positive side, receivable financing for a firm with a track record—the two most recent years being profitable—can be a relatively inexpensive finance mode that fluctuates with sales levels.

Inventory Financing

This method of financing allows you to borrow against a percentage of your inventory purchases. The collateral is still good as long as it doesn't sit in inventory too long—too long being defined by your typical selling cycle in good times. Only good inventory will be used as collateral. Like receivable financing, inventory financing can be expensive and costly administratively for risky businesses. For more proven companies, it can be a reasonably priced source of funds as your purchasing needs increase. Banks and finance companies make inventory loans.

Banks

Banks are the primary source of non-equity business financing. Banks usually require a track record and some secondary source or collateral, as well as the owner's personal guarantees. With age and a profit record of some years, the company can eventually borrow on their good name and the business' net worth without all the collateral and guarantees. As discussed earlier, banks make equipment, receivables, and inventory (tape stock and supplies) loans. Banks are regulated by the state and federal governments. The regulators second guess the bank's loans as part of making the national economy's foundations sound and to minimize depositor's insurance exposure. As a result, banks take the less risky loans with the more risky loans going to the unregulated factors and finance companies. Banks deal with more established businesses, or at least those with owners who are able to put up sizable collateral. The types of loans arranged by banks are many, and if the loan is large enough, it can be tailored to fit your particular needs. Your job then is to come up with a creative financing proposal, then run it by your friendly banker to see if it flies.

Small Business Administration

The Small Business Administration (SBA) is an agency of the federal government whose purpose is to assist the independently owned small business that cannot be financed by conventional sources. You must be turned down by two banks before applying to the SBA. A bank actually makes the loan, and the loan is guaranteed by the SBA. They generally don't approve loans to traditionally risky businesses such as films and restaurants. For more conservative ventures the SBA can be the way to go. The SBA also has special loan packages for women, minorities, and the handicapped.

THE INVESTMENT/LOAN PACKAGE

To approach a lender or investor, a loan package or investment offering needs to be prepared. Like any message, it needs to be tailored to the funding source audience. Remember you are selling both yourself and your business.

What Is to Be Financed

If you are offering stock, you need to sell your already prepared business plan (see Chapter 6) and the particulars of the stock offering. If you are financing equipment or plant expenditures, describe the planned expenditures and how they will positively affect your business. Vendors' brochures or architectural drawings give substance to a package. If you are financing working capital needs, such as receivables or inventories, give a brief history of past levels and turnaround time. Turnaround time is from the time you need financing to when you pay back the debt.

Overview of Business

Describe your business and how this particular financing fits into the overall scheme of things. A good approach is to use your annual business plan which provides a historical perspective and industry description and, incidentally, has this particular financing proposal as an integral part of the overall scheme. Remember the lender may not understand your business. Be as clear as possible.

Primary Repayment Source

How do you plan to pay back the loan or investment? For a lender, the loan payments should come from existing earnings for financed overhead items, or improved earnings for new services or lines of business financed by the loan. For an investor, she should eventually be able to get her cash out of the business—and it is hoped, at a multiple of her initial investment. Future dividends, stock offerings, or the sale of the business to another concern are typical outs for an investor.

Secondary Repayment Source

What if the primary source is a bust, how is the funder made whole? This is where home equity,

investments, a rich aunt you can con into cosigning, or your first born come in handy. The business should repay the loan, but if that fails what else do you have?

Keep the Lender Informed

Good news is easy to share, but do not let your lender or investor find out bad news from any source other than you. This applies before and after the funding. Be honest and direct with your funding sources—they represent you to their organizations and need facts to do so. You can couch bad news as positively as the situation deserves, but don't hide anything or deceive yourself to convince the lender or investor.

Everyone Is Doing It

In the *good ole days* before most of us were born, everything was on a cash and carry basis. They had prisons for debtors who fell behind in their payments. The final payment on the house mortgage was a cause for great celebration. O tempora et mores! Those were probably the days before income taxes became such a burden and the national debt such an example to follow. How times have changed and we too. Debt and interest deductions are now a way of life for us. And if inflation speeds up again, those with the most debt will be the wisest. Any loan is a risk to the lender and an obligation to the borrower. Wisely used, debt and equity financing can help you through bad times and allow your business to grow. Debt is the American way of life. Everyone's doing it.

LEASE OR BUY

When you need some equipment or a business car, you have to decide whether to lease or buy. Leases are long-term rentals. Buying or purchasing is acquiring the title as well as the use of the auto or equipment. There are pluses and minuses to each kind of transaction.

Rentals and Leasing

The discussion of leasing or buying begins with renting. Renting is obtaining the possession and use of equipment from the owner for short periods of time, usually at a fee-per-time rate. If you rent the same piece of equipment often, you should consider whether it would be financially advantageous to lease it or buy it. Would your monthly lease payments or finance payments be less than what you are paying to rent the equipment? Would your level of use likely continue through the lease term or debt payment period? Could you rent it out when you are not using it? Would buying or leasing save enough money to pay for the equipment in six months to a year, or at least during the period of probable continued usage? If you are getting the right answers to these questions, you are at the lease or buy question.

A **lease** is long-term rental. While there are no hard and fast rules, six months to a year usually separate rentals from leases. There are two generic types of leases in equipment financing. The first is an operating lease, a true lease. The second generic lease has been misnamed the capital or financial lease; in its pure form it is not a lease, but a purchase.

Operating Leases

Operating leases are true leases, in the sense of being more than a rental agreement and less than a purchase. They have a longer duration than rentals and often have vestiges of ownership. But with an operating lease, the title or ownership is not ever transferred to the person leasing the equipment, **the lessee,** from the owner, **the lessor.** Common examples of the use of operating leases are in the leasing of computers, office equipment, especially copiers, and autos and trucks. The lessor holds legal title to the equipment. He has purchased the equipment outright or had it financed. Often the lessor is responsible for maintaining the equipment in good operating condition, although this point is negotiable. Property taxes and insurance are also negotiable. Who is responsible for what depends on how much of the vestiges of ownership the lessee assumes. Leases where the lessee is responsible for the maintenance and other operating expenses, insurance and property taxes are called **net leases.** The lease payments to the lessor are net of any operating costs.

Normally, the term of an operating lease is less than the useful life of the equipment. With operating leases, the total lease payments are often less than the purchase price and finance charges of the equipment. The lease payments do not fully **amor-**tize, or pay completely, the cost of the equipment. An operating lease covers only part of both the economic life and the cost of the property. And finally, to make perfectly clear that the risks and responsibilities rest with the lessor, often the lessee under an operating lease can exercise a **cancellation clause.** Usually the cancellation clause requires some additional payments (cancellation penalties) to the lessor, but they tend not to be prohibitively costly so as to allow an out—for whatever reason—before the full term of the lease.

Tax Deductions and Credits In a straight forward operating lease deal where the ownership is with the lessor, the lessor takes depreciation and any related financing and operating expenses. The lessee is allowed to deduct his lease payments and, if the lease is a net lease, any of the operating expenses for which he is responsible, such as maintenance, insurance and property taxes. But there are many tax subtleties for more complicated situations. For instance, the lessor may not be the owner, but rather a lessee, creating a sub-lease deal. Individuals are treated differently from corporations with respect to leasing under certain circumstances. The exceptions are a little more complex, and won't be covered here. But— caveat lease complexus. Get advice if it is not a straight forward deal.

Leasing Economics The lessor expects to recoup his capital costs and financing charges from the lease payments and eventual sale of the equipment at the end of the lease or several leases. The fair market value of the equipment at the end of the lease is called **residual value.** This is what the lessor, going into the lease, thinks he can sell the equipment for at the end of the lease. The higher the estimated residual value, the lower the monthly lease payments need to be. The lower the residual value, the higher the payments need to be in order for the lessor to recoup his capital and finance costs. Where the lessor wins, is when he underestimates the residual value and sells the equipment for more than his original calculations allowed.

The lessee can obtain the use of the property with little or no capital outlay. He is not risking the full purchase price on day one. The lessee's risks are fairly controllable—he must generate enough income, or effect enough savings, to cover the monthly lease payments. If he seriously miscalculated, or technology has speeded up, he can exercise the cancellation clause. Of course, since the lessee is just renting, there is no equity build up if the equipment turns out be a long-run work horse. He'll have to buy it all over again—at a used price—if he wants to continue using it at the end of the lease term.

Off Balance Sheet Financing Operating leases are called off balance sheet financing because, since the lessee is not the owner, he can't record the property as an asset. The other side is that he doesn't record the future lease payments as a liability. It's all **off the balance sheet**—no asset, no liability. The future lease payments are required to be footnoted in the financial statements. However, many financial and credit analysts do not read the fine print, if there are even footnotes to the financials. What shows up in the financials is the monthly lease payments as an operating expense, presumedly with enough income to affect the bottom line favorably. Off balance sheet financing is usually part of the financing mix of successful companies and often financing of necessity for overly leveraged firms.

Auto Leases and Some Equipment Leases There are two formats commonly used in vehicle leasing. The terminology and concepts have also been adopted for some forms of equipment leasing.

Closed-ended lease—This is a walkaway lease. The monthly lease payments may include maintenance charges—to protect the lessor's investment, and to make life easier for the lessee. You may be liable for excessive mileage or wear and tear, but essentially this is a long-term daily rental. The lease obligations end at the end of the lease term. Drop the car off and walk away.

Open-ended lease—This format offers a purchase option at the end of the lease based on the lessor's estimated residual value of the car. At the end of the lease, you can compare the value of the car to the option price. If it's worth more than the option price, you can buy it and keep it or sell it. If it's worth less, you are liable for a maximum of three monthly payments. If the resale value is equal to the option price, you can either walk away or buy it.

Taxes and autos—The tax rules that apply to an owned auto, apply with some wrinkles, to leased autos. You must keep records of business versus personal use. The percent of business use is deductible. Like owned autos, leased cars with a fair market value over $12,800 have special luxury auto rules and IRS tables that limit the amount of lease expense write offs.

Financial Capital Leases

As mentioned earlier **financial/capital leases** are really purchases using a quasi-lease format. The lessee arranges with the manufacturer for delivery of specific equipment and then shops the banks and leasing companies for the best deal. The loan is fully amortized, that is, it is fully paid down to the financing source over the term of the lease, with interest, of course. All operating costs including maintenance, insurance and property taxes are paid for by the lessee. The lease is non-cancellable, or if cancellable, the lessee must buy out the financing source's interest in the equipment. The difference between a bank loan and a financial lease is usually the size of the down payment, with leases being smaller to none, and loans more substantial. The financial lease is really a loan with equipment as collateral. The lessee owns the equipment and the debt, and must record both on its balance sheet. At the end of the lease term the financing source releases the collateral. Because the ownership rests with the lessee in a capitalized lease, the lessee depreciates the cost of the equipment as a deduction along with the interest charges and operating costs.

When to Lease and When to Buy

A proper financial analysis would include charting the numbers, including cash flows and tax savings both ways. It is also important to add the time value aspect of money to discount it all to a present value, The calculations give you a set of answers. And, if you are dealing with reasonably predictable futures, like used cars, the answer may be right. That's the dollars and sense approach.

For better or worse, the future is generally not that predictable. In making a decision about whether to lease or buy, two factors stand out as determinant. The first is pride of ownership; the second is your ability to "out guess" the lessor's estimate of residual value.

Pride of ownership is not used merely in an emotional way. It could be that you'll take better care of a car you own, or you'll feel better about yourself and be a more effective worker. Because of who you are, ownership may make more sense financially, and it may also satisfy an emotional need. For a business, the pride argument is somewhat twisted. A key piece of equipment will often be the center piece of the business or business segment. To keep your competitive edge, the equipment needs care and continued financial investment. Would you lavish this attention on someone else's equipment? If not, you should own the equipment. The other side of this question is that it may be better to lease, especially if all you want is the equipment's utility and do not care about ownership, now or ever.

To "out guess" the lessor's estimate of residual value means you can take advantage of established trends upon which the residual values are based. For example, your use of a piece of equipment may be almost obsolescence proof. A camera that may initially be your top of the line model can be used in other aspects of the business long after its resale value has plummeted due to new technology. Or a telephone or computer system that will be technologically obsolete shortly after you buy it may continue to be a workhorse when its resale value is negligible. In these cases, it might be better to lease, since you'll probably be the only interested

party at the end of the lease, and you can get a bargain deal. Another good lease situation is one in which you think some currently needed equipment will have a short life, but the leasing world doesn't agree with you. Let them take the technological risks, and be sure to have a cancellation clause in the lease.

You call it. The important thing is first to run the numbers, and if your answers are not convincing, do a little strategic analysis. Some lease or buy calls are obvious, and others require a lucky guess.

FINANCING

There are many ways to obtain the money needed to start and run your business and to acquire the needed equipment and property to help make it grow. Become informed about your options, shop financing and alternative sources, consult professionals, and continually update your needs analysis.

PART IV
MANAGING

10 Marketing

by Ken Jurek

I'll always remember a call I received from a communications manager for one of the world's largest oil and gas companies. I had known her predecessor well. He had worked his way up and eventually out of the corporate video department and into an administrative position with the corporate parent. I remembered the time when both he and I were asked to speak on the making of our respective employee news programs. He represented the blue chip conglomerate with a slick, broadcast-oriented news show that was sent all over the world. It was produced by freelancers, hosted by professional talent, and then edited at a major post-production facility on the lastest equipment. Produced for more than 30,000 employees, this news show had received several prestigious awards. I would have given my right ear for a chance to produce just one program of that caliber and seemingly endless budget.

I, on the other hand, represented the inexpensively mounted employee news show produced for a company of about 3,500 employees. Austerity was the byline for our news program. Our anchorperson was our editor/director who also did some interviews. I did most of the camerawork, and the entire production was edited in-house on our ¾-inch edit bay.

That night, after the two of us spoke, the question and answer period revealed the interest of the crowd. Basically, they asked the big oil company video executive all about his beautifully mounted news show and neglected to ask me even a single question!

After all these years, I still envied that news program. And I wondered why his replacement was calling me.

"Well," she started, "What I need to know is: how do you justify to upper management the benefit your employees receive from watching your news program?"

I replied that I didn't have to do much justification. Once I had sold management on the concept of producing an employee news program, I continued to see what the employees felt could be improved, what they wanted to see, and then we went out and provided it to them. We simply did our own marketing and evaluation that included talking to as many people as we could. We provided an evaluation form to give us feedback on how we were doing and what we could do to make our program better.

"Yes," she continued. "But how do you get your people to take the time and watch your program?"

I indicated that I was a big believer in getting as many people on tape as possible, figuring that they would then, at least, look for themselves. Then, if we did our marketing correctly, they would like what they saw and tune in to see it again and again.

At this point I wasn't certain what information she wanted from me. She said, "Our new management team doesn't understand what we're doing with our news program. They want us to 'justify' its existence. If we can't prove that our employees are watching it and that somehow it's affecting the bottom line of the corporation in a positive manner, then we're history."

And that, in a nutshell, explains why it is important to learn marketing skills. This entire scenario could have been avoided had the big oil company video executives kept one eye on the changing marketplace while keeping the other on their product.

In a highly technology-driven industry like ours, it is easy to get caught up in all the "bells and whistles" of television production. It is difficult for us to realize that the projects, the people, and the equipment rest on our ability to market our ideas, projects and programs and to continue to market ourselves to our customers.

The problem the oil company video executive was facing was not management's inability to recognize the validity of the employee news program. Rather, it was the video department's long-term mistake in not continually going to its clients and developing new ideas, services and products.

Because the only *product* the department ever marketed was the news show, they had placed all their video eggs in one basket.

Had they developed training programs, interactive shows and other services they could have lowered their dependence on one product and spread their abilities around. In essence, what they should have been doing all along is **marketing their services.**

Whether you operate a production house or a corporate video department, you can and must market your services. With successful marketing, your department will be busier. More productions will start coming down the line. This, in turn, will raise morale in your department. This "cycle of good service" is what American industry has found to be the key to unlocking success. In *Breaking the Cycle of Failure in Services,* Leonard Schlesinger and James Heskett point out that the ability to retain clients allows companies to maintain higher profit margins. Higher profits often make for a nice place to work; employee morale is usually high.

A low turnover rate is a distinct advantage to your clients: customers like working with people they've known and trusted. Because Bill and Mary are always available to do your client's editing, your client is more satisfied than if, each time they came, a new editor was assigned to work with them. Low employee turnover leads to lower client turnover, and the cycle repeats itself.

It's been estimated that companies lose 20% of their client base each year. But, it is 70% cheaper to keep an old, established customer than it is to advertise for, court and eventually sign on and work with a new one. Because your older customers know you and your company, they are more inclined to give you more work and pay higher prices for that work because they know the kind of quality work you'll perform.

And, they'll refer other customers to you—or, if you're marketing correctly, you'll source contacts from them.

ASSESS YOUR COMPANY/DIVISION

An honest appraisal of what you do and how you do it is in order before you can start to market your services.

For example, let's say you specialize in post-production services. Ask yourself some questions:

- What are we best at?
- Why do our clients come to us? What services do they like and use the most?
- Who are our clients?
- Where do they come from?
- What do they like about us? Is it our type of equipment, services, convenience of location, friendliness and quality of our staff?
- Who are our competitors?
- What is our geographical marketplace?

The answers to these questions will tell you where you stand in your clients' eyes and in the marketplace. Once you know where you are, you'll be able to plot a course to go to where you'd like to be.

While these questions are geared toward an outside post-production facility, the same line of questioning should be used for in-house divisions. If you are an internal service division, it is important to understand that you are in the same position as all businesses. You must look at your clients, be it the vice president of public relations or the director of training, as people who can use your service.

You are in a competitive atmosphere. You have customers, you provide a product, you must answer to a bottom line.

Educational facilities often fall into the trap of thinking they have tenure for life. While it may be true that the non-profit sector is slower to react to a foundering department than the for-profit sector, departments that don't take care of business will eventually be relegated to educational television museums.

Geography is also important in helping you to determine where you stand in the marketplace. While most divisions service only their company and most of those divisions are housed under one roof or in one town, still others may attract clients from a larger area. For example, Crawford Post Production Services of Atlanta and Editel of Chicago are two post houses whose clients come from all over. That's because their reputation is good and their services are specialized enough for clients to travel to where they are.

Geography plays a role on a smaller scale as well. Years ago, I was charged with the responsibility of finding studio space for my company's in-house facility. I looked at one hundred properties and they were all turned down by management for various reasons. Then, it occurred to me that corporate headquarters wanted its new technological whiz kid to be close by. The next day I found acceptable space in the building adjoining corporate headquarters and the lease was signed in record time. Similarly, Classic Video, located in a Cleveland, OH, suburb, used to have a distance problem. Area corporate clients and advertising agencies had to travel to their facility. Surveying the situation, Classic knew that it had the right services, the right people and the right equipment. But the distance of their facility from their clients was a major stumbling block. When the company moved closer to their Cleveland clients, it enabled them to serve their clients better and removed one major obstacle they faced in their marketing efforts.

This appraisal is roughly the same for in-house departments. You should be able to identify who your customers are (i.e., sales, personnel, operations, training), who your competition is (both in and out of house) and what they like about your service. The geographic question is also important because services must be moved to where the end user is. Many video and AV departments have moved to where their corporate parent has gone.

Assuming you've done the assessment of where you and your division/company are in the marketplace, you must also assess the competition.

ASSESS THE COMPETITION

Ask yourself some hard questions regarding your real and potential competitors.

- Where else can our clients go to get done what we do for them?
- When we have lost business, to whom has it been lost? Why?
- What do our competitors have that we don't?
- What are they better at than us? Why?
- Do we have a marketplace niche, and if so, what is it?
- What do we do better than the competition?
- How do we treat our established customers?
- How do we treat potential new customers?
- Is there a difference? Why?
- How does our rate card compare with everyone else's? Do we give our customers the value they need?

Notice that the cost or chargeback of your services wasn't addressed until the end. That's because although keeping an eye on the bottom line is important, if you take care of your customers properly, having the lowest rate card in town is of less importance. It is the overall value of the services you provide that is important. It is my belief that one of the biggest reasons that corporate video/AV departments go under is because they forget to think of themselves as profit centers. The bottom line is that they need to make their corporate parent money. It can be in increased sales, increased and better communications, or better employee morale (which, as we know leads to higher productivity and better bottom lines). The department must tie itself to the bottom line and continually add to it.

DECIDE WHERE YOU WANT YOUR DEPARTMENT/COMPANY TO GO

Once you have a clear view as to where you and your company stand in the marketplace and where your competitors stand, you need to set goals and objectives for your marketing plan. Is the goal to increase or maintain business levels in the areas you presently service? Or, is it to open up new areas of business that may require investments in people and equipment?

The goal should be stated clearly in both monetary and quantitative ways. For example, a corporate video department looking to maintain its base of operations might phrase it this way:

Maintain last fiscal year's levels by producing "X" number of programs (an increase of 10% over last year). Develop three new programs or series that the department can produce. Increase the chargeback from $280,000 to $310,000.

Note: a department that stands still is actually losing ground in management's eyes.

The chargeback can be any system you use to "pay" for your department. At one company I worked at we sold our programs to individual offices to help defray the cost of producing them. Being a franchise organization, each office paid for the tapes it ordered. We even charged for our employee news program. While the cost to the offices ($50 per VHS tape) is one half what it cost us to produce a program (assuming a significant number of offices ordered the tape) that number enabled us to set a budget for the department. In essence, just like an outside production house, we were able to predict and budget. Armed with such a plan, we were able to define our objectives and keep them in line with the corporation's.

We also took in outside accounts. We used the prior year's income to help us predict what would happen in the coming year.

We found this system to be amazingly accurate. While changes in expenses and revenue have occurred, it allowed us to adjust quarterly, or even monthly. At the end of each month, my administrative assistant prepared an exhaustive profit and loss statement for the department. (See Chapter 7.) In it, we tracked how each of our *products* (our tapes) were doing, how much outside business we did and where it came from.

We kept track of our revenues on a year-to-date basis as well. This enabled us to see how we were doing according to plan and how that compared with the previous year. This was all done to make certain that, once we had defined goals for the year, we were on track to achieve those goals. We were successful in reaching our goals every year I was in charge.

THE NITTY GRITTY TRUTH ABOUT MARKETING

Dr. Charles Dygert, a marketing/sales consultant, believes that successful marketing involves the ability to "make people feel needed and important."

When Dr. Dygert explains the relationship you have with most people, he presents findings based on several psychological studies. The studies show that 25% of all people who know you care about you and like you, that another 25% don't particularly care or like you, and that the largest segment, 50%, have no feelings either way. The trick, he feels, is to convince the 50% "either way" group. By making your clients feel that they are needed by you and are important as well, you will increase you chances of being successful in your marketing endeavors.

Tom Peters, author of *In Search of Excellence* and *A Passion for Excellence*, has searched throughout industry for examples of companies that become successful by going that extra mile for their clients. It was the small company, that became successful in spite of heavy competition, that caught Peters' eye by the second book. One of his most colorful examples is that of Stew Leonard's Dairy in Connecticut. Peters found that customers were drawn to Stew Leonard's because it was a "fun and exciting" place to be. And that was because Stew had instituted several customer and employee programs designed to find out what the customer wanted and then to provide it to them in a different, humanized way.

Successful video departments are similar. They emphasize finding out what the client wants and then work to meet those needs. If you want your marketing plan to work, you'll need to get next to the customers and see things from their perspective. If they are bottom line-oriented, you'll have to convince them that you are as well. If they want creativity and digital video effects, then you'll have to show them your creative side. And to get to the client's perspective, learn the basis of great salesmanship.

MAKING YOUR CLIENTS LIKE YOU

In order to reach your goals, you must have the respect and good feelings of your clients. They must like you and want to work with you. To accomplish this, there are several helpful axioms.

1: Become Interested in Others

To have people be interested in and like you, you first must become genuinely interested in them. Clients, subordinates, bosses—all start to open up if you take the time to become interested in them. It is helpful to be interested in everyone. You never know who's going to be sitting across from you tomorrow. Former secretaries and assistants later became clients because I took the time years earlier to be kind to them. They knew that I was interested in them and they became interested in me and what I could do for them.

2: Enjoy Life

If you thing about all the people in the world that you like, you'll find that those people are pleasant, interested in you, smile and in general seem to have a positive outlook on life. Their very demeanor is upright and positive. They're fun to be around. Conversely, people who are in a terrible mood most of the time aren't any fun to be around. You tend to avoid them. There's a simple reason for this. Norman Vincent Peale, author of *The Power of Positive Thinking,* said it best when he said, "If you think life is great or lousy, you're right!" That's because what you think of life and how it treats you is directly related to how you treat others.

3: Learn and Use Your Client's Name

The person who makes it a point to learn other people's names has a natural advantage. When you use your customer's name, it immediately personalizes the conversation and gets you closer to them.

4: Learn to Listen!

It's no secret that communications skills are the number-one quality companies seek in the individuals they hire. A good communicator is worth his or her weight in platinum. A good communicator/persuader has learned to listen first to what is being said. This is not to say that good listeners are silent; what they do is listen and then ask questions to clarify what the client is saying.

When we don't listen, we shut out others people's ideas, desires, and needs. And if you're marketing your services, you want to know what they want and how they want it.

In the early 1970s, when ½-inch reel-to-reel formats and ¾-inch U-Matic VCRs became available, college instructional departments were smitten by video. I worked in such a department for a time and found that quite often we prescribed video as the solution to an instructional problem when a slide program or a set of overheads could have done just as well.

By not truly "hearing" our clients and listening to our own "videoized" brains, we forced programs onto clients who could have done without.

We also probably turned a few off to using video who found that the time involved and the expense were not what they required to get the job done.

This could happen to you. To avoid it, learn to really listen to what your clients are saying. If you're uncertain of what they want, describe it back to them in your own words so that they can clarify what they said.

5: Find Common Interests With Your Clients

Identify common interests, bring those interests up, talk about them and, hopefully, your client will feel a little better about dealing with you. You can know a little about almost everything and be prepared to mention it. If you don't know much about a subject, ask. Most clients love to talk about their favorite interests.

And don't fall into the trap of stating a view that's highly antagonistic to your client. If you should say, "Wasn't that a great election last night?" and he answers with a sneering "What are you talking about? My candidate lost!" Do what any self-respecting video marketeer should do . . . change the subject and do it fast.

6: Make Your Clients Feel Needed and Important

A client who feels needed and important around you will keep coming back to you. Be sincere at all times. But let people know how much you like them and how important their video projects are to you. They'll keep coming back. And they'll tell others about you, which is what marketing is all about.

ESTABLISHING A RATE CARD

It is important to establish a rate card, even though it may never be published or handed out. More than ever, clients, both in and out of house, want to know how much things cost. In addition, you must establish a cost system that lets you know how expensive things are to you under a variety of scenarios. Only then will you have an idea of what your services are worth and what you can charge. When establishing a rate card, you must take into account several variables.

Figure 10.1: Cost of Video by Format

Post-production Format Type	Cost
D-2	
1-inch, Betacam SP, M-II	
Betacam, M-I (RECAM)	
¾-Inch SP	
¾-Inch	
S-]VHS	
VHS	

Your Services and Their Level of Quality

Any realistic establishment of a rate card takes into account the types of services you are prepared to offer and their rankings in terms of quality. For example, Figure 10.1 shows the ranking of production levels in terms of perceived quality. One-inch Type C is better than ¾-inch SP which means that production and posting done on that format is likely to cost more. You must take into account your facility and the amount of varied services you offer. Doug Thorn, production manager for Spectrum Video of Mayfield Heights, OH, explains that "we use a lot of market research when we put together a rate card. We see what others are offering and then we try to match apples to apples with them. We compare our services in specific types of equipment and the quality of our work with our competitors."

Your Location, Physical Plant, and Employees' Commitment to Service Excellence

Once the production quality level of your services is determined, you should look at where you are located. While urban locations may mean that you are closer to your clients, suburban or even nearby rural areas can be just as advantageous, depending on the client base from which you intend to draw.

The physical disposition of your facility is also important. If your facility is stylish and people-oriented, you have the chance of getting a better return on your rates than if your edit bays are dark and unkempt.

A clear look at your people will also provide you with some idea as to how their commitment to the client affects what you can charge. Obviously, abrasive production staff, no matter how creative, will negatively impact your rate card. On the other hand, I have seen advertising agencies stand in line to pay more for "an editor they can trust who treats them well."

Your Overhead

Small businesses that are in a start-up mode often low-ball their prices in the mistaken attempt to get business. They often will charge a rate that covers materials and a little extra cost but fail to account for expenses such as overhead, utilities, rent, equipment depreciation, engineering services, tape and related supplies, people's time and profit.

Start by recording your expenses for one month. Add them up and include your complete staff's time (including you), your sales representatives and administrative office costs. Look at overtime for an average month and add that in as well. Divide those costs by however many days your facility is realistically open for business (30 days for a seven-day week, 22 for a five-day work week).

Divide this again by an eight-hour day, or how many hours you operate that you consider "prime time"; many production houses work on a ten-hour day.

You now have your cost per hour figure. This is not what you charge clients. It is only your base to break even. What you must add on is your "profit per hour"—e.g., 10% to 40%, depending on conditions. That amount will be determined to some degree by the next category.

Your Competition

"We take a look at what others are charging and take that into account when we revise our rate card. We're higher on some services and a little lower on others," explained Doug Thorn of Spectrum Video.

If your expenses are higher than what the market will bear, you may have to cut back if you are to remain competitive. All clients, internal and external, are looking for ways to hold down the bottom line.

You should consider all serious competitors' charges to help you determine the validity of your own. A serious competitor is one who offers the same or services similar to your own. The in-house production department should not attempt to justify their costs by using NBC's New York rates. Nor should it pay serious attention to a one-person operation with an edit bay in the basement; it may not offer quality video. Often, these one-person operations offer low prices coupled with limited but sometimes good-quality equipment. They can be very cost effective for clients, but they can easily get swamped. Often, they cater to only a few select clients.

Your competitors are aware of your rate structure and, as a practical business practice, you should be aware of theirs. You should also be aware of "packaging of services," which brings us to our fifth point.

Packaging Video Services to Meet Client Needs

Often, a rate card is a standard by which to go. It is only good business practice to expect that a client that would commit to 400 hours of post-production a year could command a better price for those services than one who used only a few hours of edit time a year.

The wholesale dumping of your rate card, however, to sell your services at any cost will damage your image for present and potential clients. No one respects a vendor who'll cut prices to close a deal.

But there are ways to "package" your services to meet those bottom line pressures. Certainly, the guarantee of a certain amount of business is one way. Another is to prepare a bid based on the amount of work a project is likely to involve. A third way is to group certain prices together so that one-stop shopping can benefit both you and the client: shooting, time code, graphics, effects, editing, duplication and mail services may be grouped by an hourly or package rate to accommodate the full needs of the client—and get you the business. But any packaging or coming off rate card must be looked at and included in the overall revenue scheme in order to establish a significant rate card. If 80% of your clients are working off card, that should warn you that your rate card may be too high. Conversely, if your packaging or bids continually result in getting projects with little resistance from the competition, your rates may be too low.

GETTING THE WORD OUT: ADVERTISING

When we first started taking in business from outside clients, they all had the same initial reaction. They'd walk through our online, into our studio, their jaw would drop and then they would say, "We had no idea this was up here!" Years later, this sometimes still happened. And when it happened, I cringed.

Often, I am asked how much a company should spend on advertising. This is a difficult question to answer because so many variables are involved. Your size, location and the amount and type of competition you face have a lot to do with how much advertising you should use.

I can think of few companies that advertise too much. One thing that has often impressed me is how little advertising video production companies do.

There are some companies who rarely advertise. Hershey Chocolate used to be one; Wrigley's was another. There ad budgets, even today, are relatively modest. There are others (Pepsi and Coke spring to mind) who spend a great deal on advertising and related expenses. A *Wall Street Journal* article pointed out that in an extensive double blind research taste test pitting Coke and Pepsi against local and regional colas, few people could tell the difference. Yet everyone wants the two large brands, and few people drink the lesser known colas. The conclusion of the study: advertising made the difference.

I have been a salesman all my life. Even when I headed departments that were solely in-house, I marketed by departments as a quality brand. I tried to always come up with names and logos that were easy to spot and recognize. I tried to build name recognition and make my departments' names synonymous with quality.

It always worked. Just because you are an in-house facility doesn't mean you don't have to market. Just standing pat with the clients and projects you have means that you are being left behind. Keep marketing new ideas. Keep selling yourself and your department.

That's because one of the areas communicators should know how to use is advertising. There are many ways for you to get the word out about who you are. Here are some ideas.

Direct Mail

A good return (the number of responses compared to the number of pieces sent out) on a direct mail campaign is from 1% to 2%. A great return is around 5%.

While that may not seem like a lot, direct mail is an efficient way to solicit and qualify your clients. With direct mail you can tailor your message to a specific clientele. Purchase or otherwise acquire a mailing list that includes only the criteria (i.e., region, type of organization, exact title of person who would purchase your services) you need and follow up on qualified leads that respond to your piece.

The qualified part is important because these are people who have read what you've sent and are interested. That is, they have identified themselves as either needing, interested in, or at least willing to consider your services.

An idea that future and present customers will like is the creation and delivery of a newsletter about your facility. By using a desktop publisher, it can be laid out inexpensively and provide up-to-the-minute information about what you're doing and what you're going to do. You can profile your staff (a definite morale booster!), explain new services, and provide ideas for other projects. It's a great way to "toot your horn." And everyone will read a newsletter, while most will avoid blatant advertising.

Yellow Pages

This is a viable medium that should not be overlooked. Due to the changing technology of our business, the ad should not be too specific as it may be out of date by the end of that year's directory.

A small listing or one with a few broad services should be sufficient. Remember to list all categories in which you provide services. Possible categories include: Audio-Visual Consultants; Audio-Visual Equipment-Renting & Leasing; Audio-Visual Production Services; Video Taping & Production Services; Video Tape Duplicating & Transfer Service; Motion Pictures Producers & Studios; and Communications Consultants.

A closer look at these and other titles in your area will give you an idea of your competition and suggest which categories you fit in best. You may also want to advertise in regional directories or in directories of nearby cities.

Trade Publications

National and regional trade publications offer the opportunity to reach people who not only can use your services, but who probably already do use your services. When determining whether or not you should advertise in a specific publication, determine its readership size and the type of people who read it. Determine whether these people could or would use your services.

Networking

The many meetings that professional organizations, such as the International Television Association (ITVA), Association for Multi-Image (AMI), International Association of Business Communicators (IABC), Public Relations Association of America (PRSA), and your local Ad Club, hold are great opportunities to network and tell people about your services on a person-to-person basis. Contact the national organizations to make contact with your local or regional chapters. Find out when and where the meetings are. Go to the meetings and meet people.

Hosting a meeting is a great way to strut your facilities' stuff. Advertise in the local chapter newsletter. Sponsor banquets or other functions as this helps the name of your organization while creating goodwill.

KEEPING YOUR CLIENTS

In the late 1970s, Technical Assistance Research Programs, Inc., found during a survey of customer complaint behavior that the average business never hears from 96% of its dissatisfied clients and that each of those clients told from nine to 20 others about their dissatisfaction.

The way to keep your clients can be summed up in one word—service! By servicing the account you'll assure that the lines of communication are open and that your customer will come back. If you doubt the need to service accounts, understand that the nationwide average cost for selling a new account, compared to keeping an old one, is approximately five times more expensive!

Your established clients know you and what you can do. They also know what you can't do and how far you'll go to help them in areas you normally can't provide services. For example, many departments research the best tape duplication services and either place the order or refer their clients.

It is important that you devise ways to make your clients feel comfortable after the sale of your services. In servicing clients after the initial sale, try these ideas:

• Show your clients how to maximize the service you provide. If you recorded a series of speeches for the client and the client edited the program to show to its employees, explain how the program might be re-edited to show to potential customers and others outside the company.
• Treat clients like members of an exclusive club—yours! They deserve recognition and special treatment. Offer to order lunch or dinner for them during a long edit or shoot and give them a choice of several menus.

If clients need "air" dubs, set it up so you can make them and send them overnight. Try to do everything that makes your department stand out from the rest.

Most clients prefer one-stop shopping. If that's the case, offer services that provide all-inclusive production, duplication, and distribution work.

• Review your customers' account from time to time. Even if there is no activity, your meeting with them re-establishes the goodwill you initially built and re-opens the doors. Just meeting with you may move some dormant projects ahead. I make it a point to have lunch with clients who have bought from me at least once a quarter. We may not even discuss business, but I've noticed that edit sessions and shoots are usually scheduled shortly after this get-together.
• Send a newsletter on your division or company or, at least, a letter detailing some of the things you've done in the recent past. It's just to keep them up to date.
• Send birthday, holiday or special event notes or cards.
• After each production is complete, send a thank you note. It works.
• Keep notes to help you remember spouse's or children's names. Refer to them by name in conversation or letter.
• Be a "problem solver" and an "idea maker." Don't limit your problem solving to just video; make it all of communications. Often, I'll help a client with almost any kind of problem—I even helped one get his car repaired overnight. You want people to feel comfortable about coming to you with their problems. Only then can you market your services again.
• Tell one client about a program you produced for another. This idea exchange allows them to imagine the ideas working for themselves and may even get them to think of them as programs or projects. When they take that idea to their boss or client, it often has the multiple effect of creating more work for me and my department.

CONCLUSION

Whether you are an in-house department or an outside production facility or consultant, the key to your success is your ability to market your ideas, thoughts and services. It is essential to have your clients like you.

Establishing a viable rate card requires you to look at factors within your division and outside of it. Advertising will help you get the word out on what you do, it is then up to you to capitalize on it.

And finally, keeping clients is a lot easier than finding new ones. Treat your customers like gold, with respect and kindness. Make them a part of your exclusive club. And success will be yours!

11 Personnel

by Ken Jurek

Freddy left the corporate video womb and became a communications consultant. Unlike many other freelancers, he purchased the ¾-inch production equipment from his previous employer (who had eliminated the department as a result of "downsizing").

Freddy added a client here and another one there. He took on his old concern as an outside client and charged them 10 times what they used to pay to have him in-house. Then, he came to see me.

"What am I going to do," he asked one day. "I need to hire someone, but I can't find someone with exactly the experience I want, and I can't afford the time to train someone."

Freddy had hit upon two of the main criteria for hiring. Business was good. Too good. He couldn't do all the projects he was offered. He needed to add staff. But he needed experienced staff (at a good salary) or he had to invest additional time training a person who might be more inexperienced and therefore less expensive. Eventually he hired an intern who converted to a full-time employee.

Adding staff is not to be taken lightly. It is something that should be thought about carefully. A plan should be devised that will fill the position with a person who will have the maximum impact on the organization. There are many areas to consider when adding staff.

HAVE A PLAN

Adding staff requires a carefully laid out plan. Depending on the services you offer and the budget you have, positions should be defined by job descriptions and organizational charts. When you look at your staffing plan, you need to ask yourself the following questions:

1. What must the person need to be able to do to successfully fill the position?
2. What am I prepared to pay to meet these needs?
3. Is the salary range I'm considering in line with similar positions?
4. What personality must the person have to perform this job?
5. What motivation should the person have in order to do the job?
6. To whom will the person report?
7. With whom will the person work?
8. What are those people like?
9. What will this person do every single day?

Regarding the last question, I have found it best to think about a typical day for the person who would fill the position. By thinking of the hour-by-hour duties and responsibilities, you get a "real life" view of what the person needs to be like and have in their background to successfully perform the job.

By doing this you will be able to decide what skills 1) the person must have and 2) what skills are desirable but not necessarily essential.

For example, it may be that the producer you'd like to have should be a "hands-on" type who actively participates in shoots, edit sessions, and other production functions. However, as you assign priorities to the skills you want in the new person, you may decide that a producer who is good with people and can make clients feel comfortable will meet your needs. If so, you may decide to include in your search candidates who have great interpersonal skills and less equipment experience.

BE FLEXIBLE

Many organizations inadvertently use job descriptions to restrict who can occupy the position. Use the job description as a road map. It will get you to the type of person you need to fill the position.

Like a real road map, there are many different ways to arrive at the same destination. How you get there or what the person is like or has as credentials is secondary to his or her fitting into your department.

While many organizations favor rigid descriptions, I have found success with a loose arrangement. That is, while I may have an editor-producer position available for which CMX experience is required, I may have to loosen the parameters to consider producers with some editing experience and that editing experience might not be CMX.

Therefore, for a time, other members of the staff may add editing responsibilities to their duties until the editor/producer can get up to speed. Or, they may keep the responsibilities and the editor/producer position may change to deemphasize the editing responsibilities.

Larger organizations find that "blurring" of responsibilities is difficult to manage. They prefer clear organizational charts. Smaller production departments usually have less trouble because they are already used to wearing many hats in their work.

LOOK FOR SOMEONE TO PROMOTE

There is nothing better for morale than promotions from within the organization. Executive recruiters insist that their hiring authorities first look internally for someone to promote and only go outside the organization after they exhaust this avenue.

Besides elevating morale, promoting from within eliminates the downtime a new person experiences when he or she first starts in a position. A new person has to learn a new system, new

equipment, your organizational way of doing things—and has to fit in with the rest of the staff. A promoted person already knows this.

When we had an opening for an editor, I knew that the rest of the staff didn't have the skills or desires to obtain the skills to fill the position, but I overlooked one production assistant. I overlooked him because he was relatively new and inexperienced in understanding our online editing system.

This production assistant approached me. By talking with me he demonstrated that he understood computers, that he had, in fact, created graphics on our unfriendly palette system, and that he had the creative abilities necessary to become an editor in our department. I gave him the chance to fill a limited part of the overall position and cannot believe that after all these years I overlooked someone who could be promoted!

THE TRUE COST OF EMPLOYEES

It is estimated by most personnel experts that, depending on the specific benefits plan, the total compensation received by an employee is from 30% to 40% above their salary. That means, an employee earning a salary of $20,000 per annum is often receiving $6,000–$8,000 in benefits.

Benefits packages are a double-edged sword. All organizations are driven to review their benefits packages annually to try to contain costs. But if an organization reduces its benefits package it may reduce its competitiveness in attracting new and retaining present employees.

FULL TIME OR FREELANCE

In the last decade, the pool of available quality freelance employees has risen tremendously. From time to time, especially when I've had an opening that occurred unexpectedly, I've filled it with freelance personnel.

I have also used freelance personnel to handle jobs when we have many jobs occurring at the same time or were undertaking a particularly large production.

Video departments have used freelancers to get around hiring freezes. Contract employees like freelancers and talent are usually treated as part timers thereby satisfying the freeze on permanent employment while enabling the department to complete its work.

Freelance talent usually commands a wage significantly higher than full time staff. However, that rate is all-inclusive; the freelancer has to be concerned with paying taxes and arranging his own benefits.

In a business that can experience both seasonal and project oriented downtime, the question of whether to use freelancers or permanent employees is an important one. Some points to consider: should downtime occur, a freelancer is simply out of work and incurs no further expense to the department. The permanent employee, however, is on staff regardless of the level of productivity.

Loyalty and a freelancer's popularity are also factors to be considered. Well-known and established freelancers usually have a number of clients and are often in demand. You may have a shoot you want a particular freelancer for on the 17th through 25th of the month, and they may have other jobs lined up. Some are so deluged with requests that they are reluctant to schedule too many weeks in advance. Permanent employees don't present this problem.

WHERE TO LOOK FOR QUALIFIED PERSONNEL

Should you decide to hire a permanent employee, and after disqualifying the avenue of internal promotions, you need to start the hiring process. The first part of finding the right employee for the job is knowing where to look.

Newspaper or Trade Ads: The ad should be written in clear, concise terms with an eye to selling the particular benefits of the position, department, corporation. Our industry is usually not effective in reaching qualified candidates for positions requiring a high degree of skill (i.e. editor, writer, director, producer, videographer) via newspapers. Trade journals, professional newsletters and industry job hotlines are better sources for these candidates.

Networking By using professional organizations, such as the International Television Association (ITVA), International Association of Business Communicators (IABC), Association for Multi-Image (AMI), or Public Relations Society of America (PRSA), and local advertising clubs, you can be certain that you are reaching people familiar with the skills you are seeking. Meetings, seminars and workshops also provide potential candidates.

College Internship Programs Many colleges offering communications degrees have established internship programs to provide their students with practical industry experience. These interns can provide your department with fresh personnel eager to learn. Despite interns' lack of experience, the establishment of an internship program will provide you with a stream of people and information that can benefit you when you need to hire.

Other Employers and Your Employees Other employers in similar situations can provide you with the results of their job search. Candidates that may not have been right for them may have the credentials you need. Also, your employees may provide you with names of potential hires. If you don't ask them, you may never uncover these hidden candidates. One good aspect of hiring referrals by your employees is that they may already be accepted by the staff as a known quantity.

READING RESUMES AND WATCHING RESUME REELS

Once the word gets out that you have an opening, expect a deluge of resumes and inquiries. Most of these resumes can be quickly deleted due to either the lack or irrelevance of the candidates' experience.

The reason to read resumes is to judge which candidates will progress to the interview stage. Look for "red flags" or potential problems that indicate a deficiency or problem with a candidate, including the following:

- *Long descriptions of education.* They may hide a lack of practical experience.
- *Clutter.* Resumes should be neat with lots of white space to enhance readability.
- *Length.* New college graduates should have only one page. Most of us could fit our resume on one page.
- *Qualifications.* They tend to blur the experience of the candidate. Qualifications include phrases like:
  "I have knowledge of ¾-inch SP editing . . ."
  "My exposure to studio production includes . . ."

"While with HAVSCO Productions, I assisted with . . ."

"I participated in various field productions . . ."

- *Extraneous information.* It may have been added to mask a weak resume.
- *Slickness or trendiness.* Colored paper or bizarre typefaces may hide a lack of genuine creativity or knowledge.
- *The functional resume.* A resume with no or few dates, indicating when experience was gained, could mean that the candidate operated 1-inch editing equipment years ago.
- *Equipment-heavy resumes.* The attempt to dazzle with terminology or equipment model numbers is an effort to cloud the fact that the candidate is short on experience.
- *Long work experience segments.* It may mean that the candidate has a very narrow background.
- *Sketchy background descriptions.*
- *Letters of recommendation.* Most are written by the candidate.

As you progress to the stage of viewing resume reels, keep these practical hints in mind:

1. The first three to five minutes of a resume reel should tell you all you need to know. Usually the best parts of a candidate's reel are at the start.

2. Try to ascertain exactly what the person did in each production you view. Large-scale productions can hide an individual's contribution.

3. Segments should back up the candidate's claims. If she says she is a great videographer, then the shooting, lighting and other visual components had better be great.

4. Be critical but be fair. You don't have to like the program's content to determine whether the candidate can do what he says he can.

5. Cut through glitzy, special effects-ladened productions. Ask yourself if the production works without the glitz.

6. Take notes and prepare to ask any questions that occur to you during the viewing. You may need to ask specific questions regarding what the candidates did in the various productions, how they did it, why they did it, and the thought processes involved with their decisions.

7. Be aware of the type and number of productions involved. I have viewed resume reels that were comprised of only one production; this usually indicates a lack of experience. Many productions broaden a person's experiences.

CONDUCTING SUCCESSFUL INTERVIEWS

Good hires who stay with the company and continue to contribute will reflect positively on your career. Conversely, a string of bad hires can cast doubt on your decision making abilities and reflect poorly on you.

The actual interview naturally places you at odds with the candidate. You can afford to be self-centered in your approach. The candidate, on the other hand, cannot.

Often, when dealing with creative types, finding the information necessary to make a decision is not always easy. Some candidates are natural "talkers" who can easily show their achievements. Others aren't as lucid. It may take some questioning on your part to determine their capabilities.

Employment experts have concluded that 75% of the hiring process is based on your evaluation of the person's communications skills and personality traits. We tend to hire people we like and trust. Twenty-five percent of the process is spent evaluating the candidate's background and work history.

Obviously, we hire in our own image. We tend to favor people we like. A case in point is a good-looking young man I know. He interviews well and has had three jobs in the past year so, obviously, he convinces others to hire him. But I, personally, know him to have a divisive side that hinders his ability to work with others. Because of his looks and his easy manner, people are quick to like him. It is only later that his disruptive side takes over. Yet, he continues to get hired because so much of the hiring process is based on what we think of people rather than trying to determine what they are like and how they are likely to react with others.

To insure success during the interview, there are several things you should do:

1. *Screen candidates carefully.* Have your secretary or assistant screen resumes and telephone inquires to save your time. Write a list of questions that qualify the candidates and call or have an assistant call all resumes that pass the initial scrutiny.

2. *Do your homework.* After you've set up interviews, do your homework on each before he or she arrives. Familiarize yourself with the candidate's background and plan specific questions you want to ask.

3. *Have a plan.* You should organize the questioning before the interview, write it down, then stick to your script.

4. *Ask questions in logical sequence.* Make certain that your questioning is in a logical sequence of events that flows from one area to another. An abrupt change is likely to unnerve your candidate.

5. *Take notes.* Differentiating one candidate from another is particularly difficult after several interviews.

6. *Use a conversational tone.* Practice before the interview. Ask open-ended questions that require the candidate to explain.

7. *Be businesslike.* When he or she arrives, greet the candidate and create a proper business-like atmosphere, shake hands firmly and make eye contact. It should be friendly, but restrained.

8. *Keep information confidential.* Before you start, assure the candidate of the interview's confidentiality and keep it.

9. *Let the candidate do most of the talking.* She or he is the one being interviewed. If you don't get an answer to your question, re-direct it and stay on the topic.

10. *Control the interview.* If the conversation gets off track, it is your job to ask questions to pull it back.

11. *Keep your reactions to yourself.* Avoid overreacting at all costs.

12. *Stay alert.* If the interview should drag, ask yourself whose fault it is.

13. *Write a summary.* After the interview, write a summary of it. Note your likes, dislikes, thoughts and impressions. Try to analyze the candidate's goals to see how the position you have fits him or her. It's hard to remember details when you're seeing many qualified candidates.

14. *Be thoughtful.* Avoid snap decisions; learn from your experiences.

QUESTIONS TO BE ASKED DURING THE INTERVIEW

There are four main areas of questions that make up the interview.

"Tell Me About Yourself"

This question allows you to see how well prepared the interviewee is. Candidates should be able to tell you what they have in their background that makes them especially prepared to handle the job you have available.

A second thing you'll be able to determine by asking this question is how that person will relate to others in your organization. As the candidate answers it, think about what he says and how he says it. Try to determine how the rest of your department would react to the answer. Remember, we're in communications. It starts with the individual.

Personality Questions

These questions help you determine how the candidate will fit in your organization and will give you an idea of what he is like as a person. Generally speaking, personality questions beg for a certain answer. While you may think that the desired response is obvious, pay particular attention to those who don't give it.

Some examples of personality questions are:

Are you a creative person? (Ask for some examples.)
Are you able to work under deadlines? (Ask for an example.)
Why do you think that you'll be able to do this job?
Are you able to work in pressure situations with clients?
Do you enjoy editing?
Are you proud of your work as a cameraperson?

Motive Questions

These questions help you determine if a candidate will enjoy the job and, if so, how much. Dedication in any job is a prerequisite and motive questions shed light on the candidate's reasons for being in video.

Some examples are:

Tell me, if you could have any single job, what would be the ideal job for you?
What are your strengths?
What are your weaknesses?

The Salary Question

Too often employers raise this question too early in the interview process. Inexperienced candidates are often liable to blurt out a figure that they've given very little thought to; it may have no relevance to the job. Try to determine, first, if you are interested in what the person has to offer and then, should you be interested, bring up the salary question.

Salary structures for video positions have so many variables (responsibilities, necessary equipment knowledge, geographical location, type of organization) that salary surveys offer little help in determining what you should pay. One thing is certain, however: if you offer the position to several people and they turn it down, the compensation package should be re-evaluated.

QUESTIONS TO BE AVOIDED DURING THE INTERVIEW

There is a large gray area in the interviewing process. It has to do with an individual's rights and an employer's need to know. There are subject areas that you cannot discuss.

The areas you must avoid have to do with the race, religion, color, creed and sex of the individual. Most employers afoul of the area of discrimination on the basis of sex.

Basically, courts have ruled that a prospective employer is not permitted to ask a question of one sex that would not likely be asked of the other. This means that you cannot ask questions in regard to the following:

- Marital status
- Sexual preference
- Children or the desire of people to have or not have children

Many employers don't realize that they are leaving themselves open to prosecution by asking questions about whether or not a potential employee intends to have children. You might ask the question out of genuine interest, but employers have used this knowledge to disqualify otherwise talented individuals because they didn't want to pay maternity expenses or have any downtime as a result of pregnancy.

Likewise, the marital status of a candidate may not be discussed. Employers who ask such questions leave themselves open to possible sexual harassment suits.

Additional rules and regulations may vary for each state. Be certain you know the federal and your own state laws regarding what you can and cannot ask before you start to interview.

THE GOVERNMENT AND YOUR HIRES

The small business owner may feel overwhelmed by the forms and regulations and general governmental red tape when he or she hires an employee.

Although there are variations on the regulations in each state, there are basic areas you should be concerned with when dealing with payrolls and employee benefits. Hire a good accountant and have him or her lead you through the rules and paperwork.

Areas you should cover with the accountant include:

- *Withholding taxes.* This is the amount you withhold for federal income tax purposes. The biggest mistake made by small businesses is the failure to pay these taxes quarterly. This is not "free money" for you to spend until you have to. It is part of the employee's wages and employers who are tempted to "borrow" these funds or who simply "forget" to pay them can find themselves guests of Uncle Sam in a federal prison.
- *Social Security.* This is a separate tax and, like withholding, is determined by how much is paid to the employee.
- *State and local taxes.* These, of course, vary by location. Your accountant should have a good handle on the payment of these.
- *Insurance.* Medical insurance requirements vary from state to state. You should understand what is minimally required and make your judgment as to what you can or should offer your employees. Some kind of paid participation on the part of the employee has come into vogue of late.
- *Overtime.* There are strict regulations governing whether an employee is exempt (salaried) or non-exempt (hourly with overtime). You should know when overtime kicks in.
- *Sick leave and vacation time.* Policies for both areas must be drafted with an eye to local regulations. Some states have "comp time" rules that require employers to either pay an

employee for extra hours worked or offer time off.

- *Profit sharing.* This benefit allows the employee to be rewarded by how well the company performs.
- *401 K.* A savings plan administered by the company and paid for by the employee, the company or a combination of both.
- *Bonuses.* Performance bonuses work if they are based on measurable criteria and are agreed upon by you and the employee before they are put into effect.

A *Fortune* magazine survey showed that creative compensation involving bonuses, profit sharing or stock options is available in over 70% of all American businesses. You may wish to consider your complete compensation package before you start to hire.

It is necessary to offer competitive compensation and benefits. Since you may well "get what you pay for" with respect to personnel, be prepared to emphasize the benefits of the position to each prospective employee you interview.

MANAGEMENT AND LEADERSHIP

There really is no such thing as an easy job in management. A manager is responsible for people, projects, equipment and money.

Managers tell you what to do. But effective managers do more than assign tasks and push to get things done. Effective managers are leaders. They have the ability to motivate people—they know how to inspire, how to delegate and how to lead their troops.

Leaders work to find out who does what best, assign it to them, and work with everyone involved to get things done correctly.

Obviously, being a leader is more useful than just being a manager.

There are two basic ways to increase productivity if you are a manager/leader.

Statistical process control. This is what we used to call time study, where you study a process (like an edit), break down a task into simple steps, time each one and then project how many can be done in a certain time period.

Participatory management/leadership style. Allow everyone concerned in on decision making, and take advantage of everyone's creativity.

Creative people are independent, persistent and highly motivated. They're right-brain dominant and their creativity comes out in more than just one way. They tend to be people oriented and are used to making their own decisions. They generally have high I.Q.'s. They have spent a great amount of time studying their field of expertise and are relentless in knowing all about the things they work with. (Ever ask an editor which machine he prefers and why?)

In order to get the most out of these people, you, as a leader, must strive to make them feel needed and important. By including them in the decision-making process, you are encouraging their natural creativity. This raises their confidence levels and improves their output. And it also fosters better decisions because people who contribute have a stake in the outcome.

In order to motivate others, you should be motivated within. You must be excited about your job and about what you do.

You should look for the good in people and learn to trust them. This is especially true in growing organizations where the boss has less and less contact with daily events. Hire the best, train them, work with them and then let them grow.

Shoot for the higher level in productivity, in quality, in creativity. You'll be surprised at how many times people will exceed those higher expectations.

Try to surround yourself with positive, generous, giving people. By creating a positive work environment, you'll encourage openness and tap into higher productivity. I've always strived to make my production areas like living areas.

Learn to fail. Cy Young is remembered for the incredible number of games he won—a number that's still unbroken. Often forgotten is another unbroken record of his—his career losses. He had to learn to lose in order to win. So must you.

Develop a sense of humor. Have a pep talk with yourself before you get to work. You must be "up" in order for others to be.

Learn to relax. Go for a walk. Exercise. Take real vacations where you totally get away from your work. You'll be amazed at what a happy camper you'll be if you learn to manage the stress of this very hectic industry.

TERMINATION OF EMPLOYEES

No matter how hard you try, some people don't work out. Others, who were once productive, slowly sometimes, suddenly other times, become unproductive or start to lose the inspiration they once had for the job. Other times, a downturn in business or a change in technology or the services you offer may force your hand. Someone has to go and that job falls to you.

Whatever the situation, once you determine that you must terminate an employee, there are some things you should do and others that you must do.

WHEN TERMINATION IS NECESSARY

Check Local and State Employment Laws What you can or cannot terminate an employee for varies from state to state. Check with your local state employment agencies to determine both your and your employee's rights during termination. Some states, like California, make it almost impossible to fire someone without a lot of justification. Even then, there have been court cases awarding terminated employees severance pay and damages when work was, in fact, outrageously below par.

Be certain that you don't violate an employee's rights. Stay within the bounds of the law.

Use Termination as a Last Resort Firing someone, no matter how delicately handled, always leaves a bad taste in everyone's mouth. Good hiring practices, grounded in clear evaluation procedures, should keep terminations to a minimum. But, when firing someone is necessary, it should come as no surprise to the employee. It is your duty as a manager to help employees set realistic goals. Then you should work with them to help them meet those goals. By checking their progress and offering positive criticism you will be able to judge them fairly. Should they not perform up to standards, you should tell them. If they continue to fail, termination is a realistic step that both of you will consider.

Seek Legal Advice From Corporate Attorneys They'll know governmental policies. Check on company policies as well. Many large corporations have specific procedures for firings.

Advise Your Boss It should come as no surprise to your boss. He or she should have known all along what was occurring with the employee in question. It is essential that your boss know the particulars and back you up.

While an employee shouldn't be made to feel that the entire company is against him, manage-

ment must be united. The bottom line is: your boss must back you up.

Plan the Termination Plan what you'll say, how you'll say it and give some thought as to when you're going to say it.

Phil Stella, industry consultant and former communications manager for Progressive Insurance, believes in letting an employee go on a Friday afternoon. "There are two reasons for that," he explained. "One, it's the best way to allow the employee to maintain a certain degree of dignity. If he or she has to go through the pain of packing up a desk, it's better to do it when others aren't around. Second, should a blow-up occur, it removes the chance of other employees being dragged into it."

While I've preferred Friday terminations in the past, I've started to lean away from them because an end-of-week termination allows disgruntled employees to stew all weekend. I also feel that weekday terminations give employees the workday to start looking for a new position. It keeps them busy and their mind on the process of interviewing. Whatever your inclination, plan ahead and do it according to your plan.

Termination Packages Many companies have severance packages; others develop them on an individual basis. You should examine what, if anything, should be included in the package to give to the employee after termination. Letters of recommendation, severance pay, health benefits, use of an office for a few weeks, possible freelance work with your department (many times a freelance situation works out where permanent employee status doesn't) and outplacement consulting are some of the components you may elect to offer.

Be certain that you take the time during the termination to explain the package completely. You should provide a written list to minimize misunderstandings.

Understand the Person's Stress Even if it is anticipated, terminations are hard to accept. The moment is one of extreme stress for the terminee so exercise good judgment in the handling of the situation. Tears, expressions of anger and disbelief are just a few of the feelings that can come out. Be prepared for them and handle them with dignity and understanding.

Consider a Third Party Being Present Many large corporations insist on having a personnel representative in on the termination. Should you suspect volatility in the situation, you may ask your boss or another related authority to assist. Again, the purpose here is not to "gang up" on the employee, but rather to help manage a stressful situation.

I could have used a third party when I let go two copy people. The first one handled the situation beautifully. He said he knew it was coming and that he had been looking for another job. "Great," I thought, "these guys work together and talk over everything, so the other guy should already know and be just as easy to handle."

Was I wrong!

The other man was a part-time weight lifter and when I started my speech, he exploded with anger. It was then that I realized that the only thing that stood between me and the door was him. Eventually, he left, but that impressed upon me the potential need for a third party to be near by—if only "just in case."

No Matter How Difficult, Leave the Situation on a Positive Note Video does seem like a large industry, but letting the situation degenerate into a shouting match or allowing the employee to leave on a bad note can only come back to you later.

Don't hold grudges and don't spread gossip. The purveyor of bad-mouthing is often looked down on for his indiscretions. Take the high road in all managerial situations. As a leader, people look up to you. Make sure you give them a reason to.

Learn from Your Mistakes We all make mistakes. Not all relationships work out; sometime during your managerial career you are going to be called on to let someone go. It comes with the territory.

Think about why the situation didn't work and be very concise in your assessment. If you could have done more and didn't, ask yourself why you didn't. Was it a bad hire or did the person change during the course of the job? Did you give the person a chance or were you closed-minded? What will you look for in a replacement employee to avoid the same situation occurring again.

SUMMARY

Even in an equipment intensive video department, employee salaries and benefits are usually the biggest expense you have. By hiring right, managing effectively and motivating your staff you'll do well.

Terminations do come with the job. Learn from them and go on. Life's too short not to enjoy what you do.

12 Operations

by Ken Jurek

ADMINISTRATION AND RECORD KEEPING

There are three basic ways for a small facility to set up a quality record keeping system.

1. *You or someone within your company may serve as the bookkeeper.* Spouses and relatives often assume the responsibilities in a small facility. While this is often economical, it may not be a good idea, depending on the personalities involved. This avenue can give you the greatest amount of control, but it will cost you the most in time.

2. *Business service firms.* Many of these multi-service firms are listed in the yellow pages. If you decide to do this, ask for references and seek recommendations. This avenue costs more than someone in-house with part-time bookkeeping duties, and you'll have to supply a lot more written information. These firms usually plug your information into their databases and crank out the data and records you need rather efficiently. But to operate correctly they need information from you and that is time consuming.

3. *Accountants.* Ask others who they use and try to find an accountant who is located physically close to you. Taxes can be a big headache. The various levels of government (federal, state, local, regional) change their tax rules often. An accountant or tax attorney can represent your interests and let you concentrate on running your business. In addition, never go into a tax audit without qualified representation. Always have a professional prepare your taxes. You'll save money and time and you'll sleep well at night.

Accountants, CPAs and tax attorneys also function as management consultants to your business and will provide advice in a number of other personnel and financial areas.

Naturally, you should choose the type of book-keeping resource that is easiest for you to work with and is neither too large nor too small. Small organizations may not need the services of a CPA firm, but some larger businesses may outgrow their part-time bookkeepers.

Keep Up With Your Records

It is extremely important to go over the financial records of your business each month. Whether you are in charge of a division of a large concern or the head of our own company, your financial statements, if properly designed and executed, are a working model of your business. By keeping an eye on this information, you will be able to develop strategy, make good plans for the future and take action should you see any financial clouds heading your way.

There are four important financial statements: the profit and loss, the cash flow analysis, the break-even and the balance sheet. The profit and loss statement (also known as the "P & L") and the balance sheet show the financial health of an organization (see Chapter 7).

The **profit and loss statement** subtracts expenses from income to arrive at either a profit or loss for the company. The cash flow projection details your company's cash (what is known as "liquidity") by subtracting your actual cash outlays from the cash you receive.

The **balance sheet statement** shows how those decisions have affected your business. That is, it shows the cash position or liquidity of your organization and shows what equity you have in it.

A statement that has great weight in determining the health of your organization is the **break-even statement.** It shows the volume of sales you need to balance your fixed and variable expenses. Because it shows the size of the expense "nut" you carry, it gives you a clear sales goal to minimally reach to break even. This statement helps you to make decisions to increase or decrease your expenses. Knowing your break-even goal allows you to make decisions on the leasing or purchasing of equipment, the hiring or termination of employees and the pricing of your services.

Perhaps the most important statement to manage on a regular basis is the **cash flow analysis.** Cash flow income must be greater than (or at least equal to) cash outgo. If it isn't, you won't be in business very long.

PAYABLES

Any cost, fixed or variable, is a **payable.** Salaries, rent, leases, supplies—all are the costs of operating a business. Often overlooked are related costs, including maintenance, legal, accounting, insurance, advertising, payroll taxes, delivery, equipment depreciation, and loan payments (interest and principal).

How you manage your payables has a lot to do with the credit you'll receive. Virtually every supplier—from the supplier of cardboard and plastic videotape sleeves to the manufacturer of digital video effects equipment—will ask for references from your other suppliers and is likely to run a credit check on your company. Suppliers will report how timely you are in your payments and if you've ever missed any.

Be aware of when payments are due and always try to pay on time. Most suppliers expect payment within 30 days. Usually interest charges incurred by tardy payments are spelled out on the vendor's invoice. Respect them, because those charges are enforceable in court.

Companies that are chronically late or have roller coaster cash flow problems will have trouble obtaining credit. This can cause problems, especially when you have to order $5,000 of tape stock

immediately and no one will deliver without a C.O.D. payment. Watch your payables. Chapters 11 and 7 of the Bankruptcy Code are full of organizations who either didn't or couldn't pay on time.

RECEIVABLES

Most companies operate on a 30-day payback schedule for receivables. That is, a bill isn't considered late until it is 30 days past its date of issuance. Most accountants feel that keeping most receivables, that is money owed you by clients you have performed services for, under 45 days is an acceptable practice in today's business climate. If a company takes a little longer than 30 days, but keeps it under 60 days, to pay you, then, their reasoning goes, they are still a good paying customer.

The 15 extra days, especially with large companies, is usually allotted to the invoice winding its way through the various check points to accounting (where it usually waits until the accountants decide to pay it at bill paying time).

It's not a bad idea to discuss how quickly a client normally pays *before* you begin work for them. There is tremendous reluctance to discuss anything financial with a client (including an estimate) before a production. However, by doing so you will eliminate the anguish you may have later while waiting for that check to arrive.

The initial stage of setting up a relationship with a new client is the best time to broach the payment subject. Often, just asking a simple question during the pre-production meeting such as "When we bill you for this, when can we expect payment?" will give you the information you want to know.

Get financial references from small clients and run credit checks on sole proprietors. Your banker can help. Always have a new client fill out a "credit history" form that asks for information such as credit references, nature of the business, principals involved, names, addresses, phones, banks and account numbers (you can call banks with these numbers and they'll tell you the status of the accounts), and if the company is a division of another, if it is a sole proprietorship or partnership or whatever.

It is best to bring up payment, secure a deposit, and follow regular collection procedures when extending credit to clients. To avoid problems collecting from your clients, follow these steps:

- When negotiating a production pact, commit everything to writing: who will do what, when it will be done, exact services to be rendered and the price. Copies should be signed and kept by you, the supplier, and your customer.
- If, during the production, services increase, note those changes in writing and send them to the client.
- Try to get some of the money up front. A 50% deposit is standard.
- Ask for the balance on completion. Holding the master tape is a great incentive for the customer to pay. Once you let it out the door, your leverage is gone.
- When you have balances, bill immediately and follow up with phone calls. If your past due date is 30 days, establish a tickler file to remind you to call on the 31st day.
- If you have to, be flexible. Take partial payments. And, in rare instances, discount the balance to get it immediately. $8,000 in the hand is worth $12,000 in the bush (or their bank account).
- Remember that most collection agencies won't handle accounts past 90 days. They're considered dead with little chance of collection.

Contact a few collection agencies, see what they charge and choose one to work with. You may never need it, but it's good to have one in reserve. Ask for recommendations as to how to handle clients and how far behind a client should be before you turn it over to the agency. Collection agencies work on a percentage, so if you can resolve the problem without them, you will save money by doing so. There is an amount of time that passes when the effectiveness of the agency is diminished.

If you provide production services, equipment rental or duplication services, your business should accommodate credit cards. Your local banker can help you establish the necessary processing accounts for Visa and MasterCard. Call American Express to establish an account with them. There is no fee to establish business accounts; payment is taken as a percentage of the transaction. The more transactions you do, generally, the lower the percentage you pay.

Taking credit cards as payments gives you another avenue to assure payment. An equipment rental and production staging company I know insists on a credit card number with any new account. They explain that, of course, they will be glad to bill whatever company the person represents. But, just in case, they'll need the card "as a backup." Then, they explain, if the bill isn't paid, it will be charged to the prearranged credit card account.

PERSONNEL RECORD KEEPING

A personnel file must be kept on every employee. Prepare a basic file folder with most of the information you should have: when the employee started, their compensation, any raises, their W-4 and I-9 forms, any benefit cards or data sheets they need to fill out, their marital status, social security number, address, next of kin, phone numbers and age.

Fill out an employee status change form each time you give a raise, change duties and/or job title, or terminate an employee. It should be dated and, in a large organization, be signed by at least two immediate supervisors.

Naturally, any employee reviews, recommendations or discipline write ups should be included in the file. In this era of litigation, everything that can be written down, should be. Any incident, good or bad, should be recorded.

There are three good reasons for keeping up-to-date personnel files. It gives the employee and his or her supervisors an honest record of their status within the company; it could be pivotal in any legal action taken either against the company by the employee or vice versa; and it helps when giving recommendations on past employees to future employers.

This last reason can be a "sticky wicket" for employers. Courts have ruled against employers denying past employees their right to get a job in the field of their choice by the passing of false information or rumors. However, they have upheld and even admonished past employers who withheld documented information that put the past employee in an unfavorable light.

A graphic artist I had let go a year earlier asked me for a recommendation. I was astonished that he would list me as a recommendation, but after the outside reference checking firm contacted me I realized what happened. The artist had listed a "friendly" client of his within our company as his supervisor. But the client hadn't gone along with it and had told the truth to the checking firm, which contacted our personnel department and was told that I was, in fact, his former supervisor.

Things started to boil when the firm contacted me while I was on a business trip out of state. The artist contacted me the minute I stepped back into my office and asked for "my help in getting this job."

Now, I am the most sympathetic person when it comes to career development. But, I will not lie. Honesty and integrity are the cornerstones of my business and personal life. And, Ivan Boesky and his ilk aside, I think you will find that most business people are scrupulously honest. It's the few bad apples that ruin it for the rest of us.

The artist had three "white lies" he wanted me to make. I explained that I would try to help, but I couldn't and wouldn't lie.

I pulled his personnel file which had detailed our past relationship with him. Checking with legal, I stated what was written and took care not to embellish. The obligation to stick with the truth was made easier by using detailed personnel files.

COMPANY POLICY AND PROCEDURES

Company policy is enforceable *if it is written down and handed out to all affected employees.*

Some companies have elaborate codes that cover everything from benefits to employee rights to termination rules. Others post a few memos outlining vacation and sick time policy and little else. There are advantages to both.

When I was a tenured professor, the college had a coded policy manual that was as thick as the Manhattan phone directory. The encoded procedures made certain that everyone was treated uniformly and allowed supervisors to manage by rote. But it was cumbersome and more often than not employees' special situations were negated. The spirit of the rules was lost in the need to treat every situation exactly the same.

If you manage people, you know that nothing is constant or uniform and that every situation is different. For example, your company's vacation policy may allot two weeks after an employee has worked there a year. But what do you do about the employee you hire in September who tells you about the family reunion in Norway she *has* to

attend next July? The cookie-cutter approach dictated by elaborate company policy procedures will penalize this employee.

Large concerns need exacting policies and procedures. Basically, smaller companies will do well with a more flexible approach. This flexibility can be one of the charming things of working for a small company. The bigger the company gets, the more policy has to be written, and the less flexibility there is. Finding a balance is the trick.

VENDOR/CLIENT RELATIONSHIPS

Imagine a company where its customers like to go because "it's so much fun to be there!" It's not too hard to imagine because the move to make customers a part of your business is becoming more evident each day.

There has been a tremendous move in the service industries of America (and video *is* a service industry) to shift priorities and reemphasize the role of the customer. Part Japanese influence, part increased competition—companies are learning to put, as industry guru Tom Peters once said, the customer first, second and third in terms of priorities.

By making your customers feel needed and important, you will develop a positive business cycle throughout your organization. It's simple. The cycle of providing good service to your customers raises their satisfaction level and thereby lowers your prospects of customer turnover. It is estimated that it is 70% more expensive to find a new customer as it is to retain an old one.

By lowering customer turnover costs, you'll enjoy higher profit margins, thereby increasing employee satisfaction. Employees working with the same customers over a period of time often develop relationships that blossom and help your business. Employees then feel needed and important to your company and that, in turn, makes your customers feel positive and the cycle of good service starts over again.

INCREASING EMPLOYEE MOTIVATION AND CUSTOMER SATISFACTION

There are some definite methods to make your customers and employees more a part of your business:

Allow employees to "own" a job. In non-broadcast video, we have the luxury of taking certain projects and running with them from the idea stage to scripting, through production into post and out to distribution. This enables personnel to become involved and bear direct responsibility for a project's success.

Arrange flexible working hours. Multiple post-production shifts can allow flexibility. Freelancers, college students and parents of small children can benefit from a flexible schedule that allows them to work while maintaining other activities.

You as an owner must take the lead and be totally committed to your customers. Your employees will follow your lead. If you act as if your customers were a bunch of idiots, that attitude will spread and you'll be finding work hard to come by.

Try to hire people who genuinely like and are interested in others. These people will take an active interest in your customer and that means increased business for you.

Try to grow your own trainers. Have senior employees be mentors to junior or new employees. They'll be able to get them up to speed while impressing upon them the philosophy of your organization.

Try to remain focused on retaining your customers. Not only is it cheaper in sales and advertising expenses, but established customers tend to buy more, refer new clients, and often are willing to pay more for your service because they know you.

Try to put yourself in their place! Think of yourself as a customer of your company; then, think of yourself as an employee. How would you want to be treated? How are you treated now? Are there any significant differences?

Link employee compensation and benefits to customer retention. By being willing to pay for extra service, you're sending your employees the message that the customer is, indeed, number one.

Train your employees in proper phone etiquette. With cellular portable and car phone use growing at a tremendous rate, more business than ever is being conducted on the telephone. Most phone companies offer booklets on Phone Power. Have your people identify who they are each time they pick up the phone. It allows the customer to know who they are. It allows them to identify the person and realize that they are talking to a person, not just a voice. It immediately makes your organization more customer oriented.

Finally, spend more time with your customers and employees. Learn to view things through their eyes. By looking at productions and projects through your client's vision, you'll be able to see what their goals are and what shortcomings they have that you can help.

OBTAINING BUSINESS INSURANCE

There is no such thing as too much insurance. By insuring your business against the various perils that exist, you are insulating yourself from trouble.

There are three basic types of insurance you must have to operate a video production company. The first is facility insurance. You should have insurance on the building you are housed in if you own it. If you are operating out of your home, you should explore obtaining additional insurance for the dedicated business part.

Whether you own or rent your facility, the second type of insurance you will need is content insurance on all the equipment you own, lease, or rent. Remember that each piece of equipment should be itemized and scheduled separately. Update this list as you add new equipment and delete older gear.

The third type of insurance is also the most important in today's America and the one we often forget. Business liability insurance is essential to cover lawsuits arising from anyone who thinks your services, people or products have harmed them.

For example, you could produce a commercial for a product that caused problems and be added to the lawsuit as one of the defendants based on the fact that you produced the commercial. You could work on a script, have an actor deviate from it and slander someone and you become liable for it. Someone could claim that your EFP truck ran over his bushes or dog during a shoot. Or someone could claim that she was physically shocked when she touched some of your cables during a shoot.

You needn't be guilty of any wrongdoing. It is simply enough for people to claim that they were harmed to put you at risk. Liability insurance should be looked at as a way to protect yourself and your organization. I recommend getting a one-million to three-million dollar umbrella policy.

In addition to these three types of insurance, there are various other types you should explore. Ask yourself if you fit into any of these scenarios. If you do, consider buying insurance now:

Performance insurance. You're producing a program with certain actors. What if they don't show up or fall ill? You are producing a car race or something that occurs outside. What happens if it rains?

Contractor's liability. You sign contracts to write, produce or direct, but for reasons beyond your control, you can't produce. Now what?

Workmen's compensation and related disability insurance. During a crane shot, your cameraman falls and breaks several bones. During his rehabilitation, who's going to pay?

Health benefits. Your graphic artist develops a tumor that, when removed in the hospital, costs about half of what he makes in a year. Who's going to pay?

Bonding and money handling insurance. Your retail division handles lots of cash and credit card sales. One day, all the money is missing. Who's going to pay?

OBTAINING COPYRIGHTS ON NEW PRODUCTIONS

Any work (i.e. script, video or audio production, graphic, photo) that you or your company personally creates and produces carries a copyright the instant it's finished. Under the revised copyright law of 1978, if you meet the criteria of authorship (it is original, it is your product, etc.), it is *automatically* considered copyrighted when you create it. Since most video work is prepared over a period of time, the exact date of copyright is when the program, script or series is completely finished.

All original work should carry the copyright symbol, the year of copyright and the owner of the copyright's name. For example, on all video productions, labels and related written materials for a training tape program, you might have the following listed:

©1990 Joe Bob Productions, Inc.

To register the copyright so that you can protect your work, you must apply to the Library of Congress. The registration fee is $20. However, rather than file it yourself, you'll find it better to have a patent and copyright attorney do it for you. It will cost a little more, but then you know each step will be followed correctly and that the copyright is properly registered.

For up-to-date information regarding the registering of copyrights, contact the Copyright Office, Library of Congress in Washington, DC, 20559. When you register your work, you'll need to send the following three elements in the same envelope:

1. A completed registration form.
2. The fee for each application.
3. A copy of the work being registered.

Generally, you will want to submit a completed program with its related written training and/or promotional materials. This demonstrates that the program is intended for circulation and is, in fact, published and in circulation.

THE USE OF COPYRIGHTED MATERIAL WITHIN YOUR PRODUCTIONS

The law is quite strict about copying materials. There is a body of legal cases related to the unlawful duplication of books, magazines and scores of music.

Every month there are clients who want to use the theme from "Rocky" or "Flashdance" or clips from NFL Football or some other commercial or public television program. I tell them that *it is illegal to use copyrighted material without the copyright holder's permission. You could be sued and held liable. Don't do it!*

Use music, video and film libraries from whom you can purchase licenses for single or multiple usage. If you have to have a particular theme or movies scene, there are a number of clearing agents who, for a fee, will contact the copyright holder to clear it for you. Most of them are located in New York City or Los Angeles.

The chances are, if you use copyrighted music or film and it is used for a private organization on a very limited basis, you probably won't get caught. But that's the same philosophy most drunk drivers subscribe to when they get behind the wheel.

What would you rather do?

THE UNFORTUNATE REALITY OF THE COPYRIGHT LAW

Similar to copyright abuse is the unlawful *copying* of video programs. We all have VCRs in our homes and we all tape television programs. Many of us have two VCRs so we can copy favorite motion pictures. And everyone seems to dub audio tapes of their favorite discs. While it is not legal to do so, there is very little anyone, including the copyright holders care to do about it. Until you try to sell the copy.

If you sell copies of copyrighted material, you could have triple damages assessed against you. This is great news for the copyright holders. Unfortunately, proving that someone has illegally copied and sold your program is difficult and expensive.

In a case I was involved in, we received information that someone was duping and selling copies of a video training series we had produced for a client and were marketing nationally. We were selling the eight volume series for $695 and he was duping two programs on one tape and selling the resulting four tapes for $150.

We arranged for another party to contact the illegal duper and purchase the tapes with our money. In addition to the illegal dupes, our contact even got a signed receipt.

Armed with this information, we got our lawyers into action. The case never went to court; it's doubtful it would have made it to court any time soon. We settled out of court for a decent sum and a public apology. But along the way, we learned something very significant.

Had our illegal duper not sold his copies, but had given them away, we could not have expected much in the way of damages. That's just the way the law is written. Fortunately for us, Mr. Greedy had to sell the tapes thereby exposing himself to triple damages and making his exposure significant.

DAILY OPERATIONS

There are three forms of communication you must have to stay in contact with your clients. Essential to your business health are the mail, with its various local courier and overnight services, electronic facsimile, and the telephone.

Between clients and vendors, there are some guidelines to follow regarding communications:

Courier services: If you deal with ad agencies and corporations, familiarize yourself with your area courier services. Promptness and neatness in scripts and proposals is almost as important as creativity and pricing. If you don't know which service to use, ask your clients for recommendations. Being quick and efficient impresses on clients that you value their business.

Most courier services will quickly familiarize themselves with you, your business, and the types of things you're likely to send. That's important because it's hard to carry posters on a bike.

Overnight delivery services: There are about three or four large competitors. If you do enough business, they may give you volume discounts and a set pickup time when they'll come to your facility each day.

The United States Postal Service and the United Parcel Service. If your mail is heavy or if the number of packages you send is astronomical, you can probably get pick up and delivery on UPS. Familiarize yourself with pickup times and, if you need to get it in the day's mail, have it ready on time.

Facsimile machines: There's no sense fighting it, virtually every company, service organization and consultant has a fax. You may want to get double duty out of your computer by adding a fax board to your word processor. Shop around since the field is quite competitive. A small fax machine (but usually not the smallest) may be all you need.

Go the extra mile and put your facsimile machine on a dedicated line. That way your company is always reachable. I have received production bids via fax from companies I didn't know existed because they saw my fax number in an area production guide.

Telephone systems: This has become a highly competitive field with an endless supply of vendors hawking their wares. When planning on purchasing or leasing a phone system, ascertain your needs, allow for growth, be practical but don't go crazy with the options. Be sure to sign with a quality long-distance service and get calling cards for your key personnel and sales people.

Look at your written communicative services and make sure they are top notch in every way. Like the salesman with scruffed shoes, you don't want your message destroyed by your medium. When mailing or sending tapes, graphics, storyboards or slides, utilize the medium that will get it there at least two steps ahead of what your client expects.

That way you'll be one step ahead of your competition.

EQUIPMENT MAINTENANCE

In video it's important to be quality oriented. Although a few companies can make their living being quick and dirty, poor looking video with low quality will last only so long. When clients find that there is better quality readily available, they'll make their move to it.

Talk with your equipment vendor or local video engineering company and devise a regular maintenance schedule for your equipment. Like a Mazeratti, your video facility is a finely tuned instrument that requires constant attention and tweaking.

It should be someone's responsibility within your organization to constantly check the equipment and perform simple testing on it. Depending on the size of your organization, you might have someone whose job it is to perform minor maintenance.

Don't fool yourself into thinking that you are saving money by stretching the amount of time between regular checkups. It would be insane not to have your car's oil changed every 5,000 miles; likewise, it would be ridiculous not to have your VCRs overhauled after 1,000 hours. You are running the risk of ruining your equipment.

For a small facility with mainly ENG equipment and simple cuts-only editing, you should be able to get along with periodic maintenance checks by your vendor's engineering department. For larger systems (such as a post facility with digital effects, graphic boxes and studio cameras), you should probably contract with an engineering consulting firm to provide regular and emergency maintenance.

For larger production facilities with multiple formats, several post-production bays, complex graphics equipment and duplication facilities, economics dictate on-staff engineering.

On all video maintenance—you get what you pay for. You may be able to strike a bargain by contracting for a year of services and you may be able to get package deals when repairing several similar units. When dealing with manufacturers, walk softly, but carry a big stick. Most want to provide good service. By complaining loudly and getting the ear of vice presidents, you'll find manufacturers to be very receptive.

When planning a facility, it is essential that you obtain the services of a video engineering consultant group early. Their input is also required for any major moving of equipment and redesign of your present facility.

When building, your video engineers should be totally responsible for the building. A friend of mine showed up at his new building to discover three new air conditioning units atop his studio. When fired up, the vibrations shook the building to its foundations.

When building our new studio, I found that a four-foot tall heating duct was ten feet from the wall and eight feet from the ceiling. The builders couldn't comprehend the problem, but after my engineering consultants showed up, the duct was moved.

And that's why you need a video studio expert when you're building or rebuilding a facility.

CONCLUSION

The operation of your video organization depends on one person—you! You alone are able to keep things organized and on target. Your staff looks to you for guidance and ideas. Your organization often bears a direct resemblance to you. Make it look and operate efficiently, and the reflection will be completely on you and your staff!

Index

About the Authors

Scott Carlberg is an industrial relations and human resources specialist for Phillips Petroleum Company. He also performs a variety of community relations and press outreach activities. Previously, Mr. Carlberg handled management and video communications for Phillips, including the management of Phillips' internal video facility which served a worldwide audience of Phillips employees.

Mr. Carlberg has written and lectured on numerous topics in video and management, and is author of the book, *Corporate Video Survival: A Book of Strategies,* published by Knowledge Industry Publications, Inc.

Ken Jurek has been in corporate, institutional and educational video for more than 20 years. Currently, he is a visiting professor at the University of Akron's School of Communication. Previously, Mr. Jurek spent 10 years heading up the television, graphics and print departments for one of the nation's largest executive search firms. He was also an associate professor at John Carroll University and a professor and department head in audio visual technology at Cuyahoga Community College.

Mr. Jurek writes for a number of business and trade publications, and is the author of *Careers in Video,* published by Knowledge Industry Publications, Inc.

James H. Spalding Jr. has more than 20 years of professional, managerial and administrative experience in production, distribution broadcast and media communication business. As principal of Spalding & Company, he provides tax, financial and management consulting services to many businesses and professionals in film and video. Mr. Spalding writes a column on these subjects for the monthly publication *Film/Tape World.*

Mr. Spalding was formerly vice-president and chief financial officer of One Pass Film and Video (now Editel). He was also chief financial officer of KQED, Inc. and audit supervisor of Arthur Young & Company.

Neil R. Heller is president of the DJM Group (Twin Peaks, CA), advertising and public relations, technical product development, marketing, and product manufacturing company. Mr. Heller's work has focused on the start-up of new technology, and he has written numerous articles about the practical applications of technology for broadcast, security, graphics and animation. His book *Understanding Video Equipment* was published by Knowledge Industry Publications, Inc. Mr. Heller has been a product manager for several companies.